GCSE
IN A
WEEK

Biology

Hannah Kingston and Kerry Duffield

Revision Planner

How Science Works

Science attempts to explain the world we live in.

The Thinking Behind Science

Scientists carry out investigations and collect evidence in order to explain phenomena and to solve problems. Scientific knowledge and understanding can lead to the development of new technologies that have an impact on society and the environment.

Evidence

Scientific evidence provides facts that help to answer a specific question and either support or disprove an idea or theory. Evidence is based on data that has been collected through observations and measurements.

Evidence should be:

- **reliable** (if you do it again you get the same result)
- **accurate** (close to the true value)
- **valid** (must be repeatable, reproducible and answer the question).

Observations

We can use our existing models and ideas to suggest why something happens. This is called a **hypothesis**. We can use this hypothesis to make a **prediction** that can be tested. When the data is collected, if it does not back up our original models and ideas we need to check that the data is valid, and if it is, we need to go back and change our original models and ideas.

Variables

- An **independent** variable is the variable we choose to change to see what happens.
- A **dependent** variable is the variable we measure.
- A **continuous** variable (e.g. time or mass) can have any numerical value.
- An **ordered** variable (e.g. small, medium or large) can be listed in order.
- A **discrete** variable can have any value that is a whole number, e.g. 1, 2.
- A **categoric** variable is a variable that can be labelled, e.g. red, blue.

Line graphs are used to present data where the independent variable and the dependent variable are both continuous. A line of best fit can be used to show the relationship between variables.

Bar graphs are used to present data when the independent variable is categoric and the dependent variable is continuous.

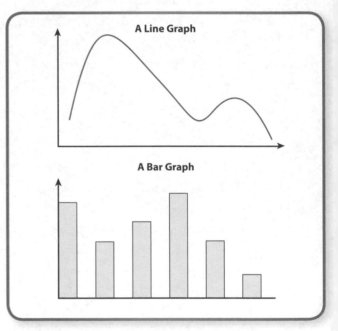

A Line Graph

A Bar Graph

Controlling Variables

In a **fair test**, the only factor that should affect the dependent variable is the independent variable. Other outside variables that could influence the results are kept the same (**control variables**) or eliminated.

Control groups are often used in biological and medical research to make sure that any observed results are due to changes in the independent variable only.

Scientists often try to find links between variables.

Links can be:

- **causal** – a change in one variable produces a change in the other variable
- a **chance** occurrence
- due to an **association**, where both of the observed variables are linked to a third variable.

Measurements

A number of factors can affect the reliability and validity of measurements, including the accuracy of measurements, the resolution of instruments, or human, systematic and random error.

Any **anomalous** (irregular) values should be examined to determine why they appear.

The data collected must be **precise** enough to form a **valid** conclusion: it should provide clear evidence for or against the hypothesis.

Science in Society

Sometimes scientists investigate subjects that have social consequences, e.g. food safety. When this happens, decisions may be based on a combination of the evidence and other factors, such as bias or political considerations.

The reliability of an investigation can be increased by looking at data from **secondary sources**, using an alternative method, or the **peer review** process where qualified professionals check an individual's work.

Science can raise certain issues involving **society**, **economics**, the **environment** and **ethics**.

Although science is helping us to understand more about our world there are still some questions that we cannot answer, such as 'Is there life on other planets?'. Some questions are for everyone in society to answer, not just scientists, such as 'Should we clone humans?'.

QUESTIONS

QUICK TEST

1. What is an independent variable?

2. What is a dependent variable?

3. What is an anomalous result?

4. What is an ordered variable?

5. How would you present data that involves a discrete variable?

EXAM PRACTICE

1. A student investigated the amount of water lost by a plant during different times of the day. Their results are shown in the table below.

Time	Water lost (mm³)	Time	Water lost (mm³)
00:00	6	12:00	248
02:00	5	14:00	230
04:00	7	16:00	128
06:00	55	18:00	112
08:00	125	20:00	8
10:00	105	22:00	7

a) What is the independent variable? **(1 mark)**

b) What is the dependent variable? **(1 mark)**

c) Identify any anomalous results in this investigation. Why do you think they are anomalous? **(2 marks)**

d) At what time of day is water lost most quickly? **(1 mark)**

e) In total, how much water did the plant lose during the day? **(1 mark)**

f) How could the student make their results more reliable? **(1 mark)**

g) What type of graph would you use to present these results? **(1 mark)**

Healthy Living

To be healthy, you need to eat a balanced diet and take regular exercise.

It is important to balance energy intake with the amount of energy used up in exercise. Otherwise you could become overweight or underweight and suffer from health problems.

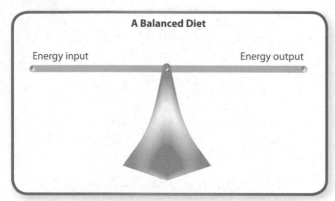

A Balanced Diet

Energy input | Energy output

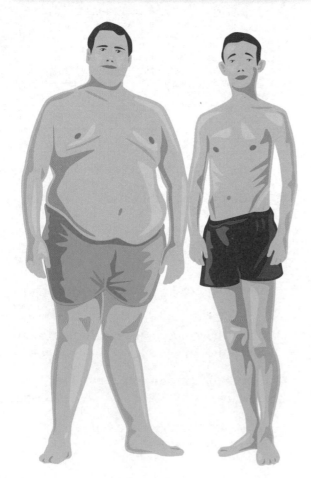

A balanced diet consists of:

- **carbohydrates**, which provide sugar and starch for energy
- **proteins**, which are made up of amino acids and are needed for growth, repair and replacement of cells
- **fats** (made up of fatty acids and glycerol), which are needed for cell membranes, insulation and producing energy
- **vitamins and minerals**, which are essential in small amounts (e.g. iron and Vitamin C)
- **fibre**, which keeps food moving through your system and prevents constipation
- **water**, which is essential for all the body's chemical reactions.

Deficiency Diseases

A person is **malnourished** if their diet is not balanced. This may lead to a person being overweight or underweight. An unbalanced diet may also lead to deficiency diseases or conditions such as Type 2 diabetes.

Kwashiorkor is a protein-deficiency disorder that is common in developing countries because the diets consist of mainly starchy vegetables. Overpopulation and limited investment in agriculture also contributes.

A person loses mass when the energy content of the food taken in is less than the amount of energy expended by the body. Exercise increases the amount of energy expended by the body.

Being overweight (**obesity**) is linked to increased health risks such as arthritis, heart disease, Type 2 diabetes and breast cancer.

Low self-esteem, poor self-image and desire for perfection can lead to a poor diet and increased health risks.

Metabolism

The rate at which chemical reactions in the cells of the body are carried out is called the **metabolic rate**. This varies with the amount of activity you do and the proportion of muscle to fat in your body. Metabolic rate may also be affected by your genes.

Cholesterol

Cholesterol is made in the liver and is found in the blood. Eating saturated fats can increase blood cholesterol. Cholesterol can restrict or block blood flow in arteries by forming plaques, leading to atherosclerosis. Genetic factors, smoking and alcohol can also contribute to the effects of cholesterol and increase the risk of heart disease.

Statins are drugs that can be used to lower levels of cholesterol in the blood; they are used to lower the risk of atherosclerosis, heart and circulatory disease.

Calculations

The estimated average daily requirement (**EAR**) for protein varies depending on age and if you are pregnant or breastfeeding. It is calculated using the following formula:

$$EAR \text{ (g)} = 0.6 \times \text{body mass (kg)}$$

BMI means **body mass index**. It is calculated using the following formula:

$$BMI = \frac{\text{mass in kg}}{(\text{height in m})^2}$$

A BMI of less than 19 = underweight, 19–25 = average weight, 26–30 = overweight and over 30 = obese.

Health Claims

Some food manufacturers claim that particular foods benefit our health. For example, manufacturers of probiotics containing *Bifidobacteria* and lactic acid bacteria *Lactobacillus* claim they improve digestion.

Prebiotic oligosaccharides are a form of carbohydrate that are said to encourage the growth of bacteria such as those found in the gut, which improves digestion.

Plant stanol esters contained in products such as spreads and yoghurts are claimed to reduce cholesterol levels.

SUMMARY

- If a person's diet is not balanced they are malnourished.

- The metabolic rate is the rate at which chemical reactions in the cells of the body are carried out.

- A healthy body mass index (BMI) is between 19 and 25.

- High cholesterol can cause many health problems.

QUESTIONS

QUICK TEST

1. List all of the components of a balanced diet.

2. How do you calculate BMI?

3. What drugs are used to lower cholesterol levels?

EXAM PRACTICE

1. Which type of fat increases blood cholesterol?
   (1 mark)

2. What does the term 'metabolic rate' mean?
   (1 mark)

3. Describe why fats are important in our diet.
   (3 marks)

Tissues, Organs and Organ Systems

Cells differentiate and become adapted for specific functions.

Tissues

A **tissue** is a group of cells with similar structure and function. For example:

- Muscular tissue that contracts to bring about movement.
- Glandular tissue that can produce substances such as enzymes and hormones.
- Epithelial tissue that covers some parts of the body.

Organs

Organs are made of tissues. One organ may contain several tissues. The stomach is an organ that contains the following:

- Muscular tissue that churns the contents.
- Glandular tissue that produces digestive juices.
- Epithelial tissue that covers the outside and inside of the stomach.

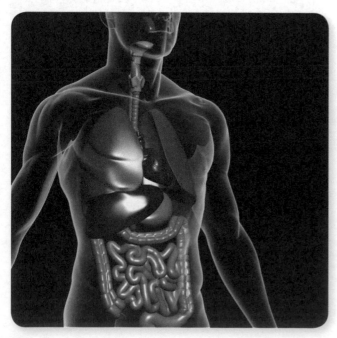

Organ Donation

Organs can be donated by living or dead **donors**. The supply of organs is limited by a shortage of donors, correct tissue matching, size and age. One problem with organ transplants is potential rejection by the immune system. Also, recipients need to take immunosuppressive drugs for the rest of their lives to prevent rejection.

Potential donors can sign on to a register of donors. There are many ethical issues concerning organ donation. Some people believe that it is wrong to remove organs from a person. Others believe that everyone has a right to life. Should an alcoholic be given a liver transplant? What about supplying the clinically obese with a heart transplant? Governments have set out codes of ethics to help people make decisions.

Some organs can be replaced by mechanical aids. The use of these aids is restricted by size, an adequate power supply, the types of materials used and how the body reacts.

Organ Systems

Organ systems are groups of organs that carry out a particular function.

For example, the digestive system includes the following:

- **Mouth** – physical digestion occurs by chewing.
- **Salivary glands** – produce amylase enzymes to begin chemical digestion of carbohydrates.
- **Oesophagus** – has circular muscles in its wall that contract and squeeze behind food to push it along in a process called **peristalsis**.
- **Stomach** – produces hydrochloric acid and protease enzymes (which work better in acidic conditions). Mechanical and chemical digestion take place here.
- **Pancreas** – produces lipase, protease and carbohydrase enzymes.

- **Small intestine** – more chemical digestion and absorption of soluble food occurs.
- **Liver** – produces bile that neutralises acid, providing alkaline conditions for the enzymes in the small intestine to work more efficiently, and also improves fat digestion by emulsifying fats.
- **Gall bladder** – stores bile before it is released into the small intestine.
- **Large intestine** – where water is absorbed from undigested food producing faeces.
- **Rectum**
- **Anus**

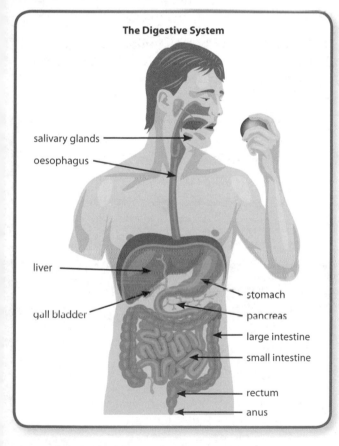

The Digestive System

salivary glands
oesophagus
liver
gall bladder
stomach
pancreas
large intestine
small intestine
rectum
anus

Physical digestion is the physical breaking down of food by a mechanical process such as teeth chewing food. This allows substances to pass through the digestive system more easily and provides a larger surface area for enzymes, involved in chemical digestion, to work with.

SUMMARY

- Similar specialised cells working together form tissues, which form organs when they work together and organ systems when organs work together.
- Organ transplantation is limited and has ethical issues.
- The digestive system is involved in breaking large insoluble food into small soluble food allowing absorption into the blood.

QUESTIONS

QUICK TEST

1. Define the term tissue.

2. Name three types of tissue found in the stomach.

3. What types of enzymes does the pancreas produce?

EXAM PRACTICE

1. Place these parts of the digestive system in the order they are met by food passing through them. **(1 mark)**

   oesophagus   mouth   small intestine

   anus   stomach   large intestine

2. Give two roles that bile plays in digestion. **(2 marks)**

3. Give three reasons why the supply of organs for transplants is limited. **(3 marks)**

4. Following an organ transplant, why are drugs given to the patient to suppress the immune system? **(2 marks)**

Proteins and Enzymes

Proteins have many functions both inside and outside the cells of living organisms. Proteins such as enzymes are used in the home and industry.

What are Proteins?

Proteins are long chains of amino acids, folded to produce a specific shape. Proteins play important structural roles such as muscle tissue and collagen. Hormones, antibodies and enzymes are all different types of protein.

Proteins are used as a source of energy when fats and carbohydrates are unavailable. Proteins are not stored in the body.

Proteins that come from animals are called 'first class proteins' because they contain all essential amino acids (these cannot be made by the body). Plant proteins are called 'second class proteins'.

What are Enzymes?

Enzymes are proteins produced by living cells. Each enzyme has its own number and sequence of amino acids resulting in specific shapes and functions.

An enzyme has a specific shape that allows other molecules, called **substrates**, to fit into its **active site**. The enzyme and the substrate fit together using a **lock and key** mechanism. Enzymes are highly specific to substrates.

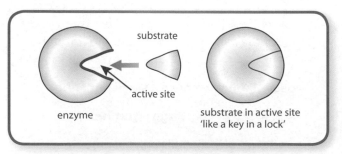

substrate

active site

enzyme

substrate in active site 'like a key in a lock'

Enzymes are called **biological catalysts**. They speed up or catalyse biological reactions inside and outside living cells, including respiration, DNA replication, protein synthesis, photosynthesis and digestion.

Enzyme activity is affected by pH, substrate concentration and temperature. Each enzyme has its own **optimum conditions**, e.g. pepsin in the stomach works in acidic conditions.

Enzymes in the human body work best at 37°C which is our normal body temperature. At too low temperatures, the rate at which the substrate joins with the enzyme's active site is slowed down so the reaction is slower. At too high temperatures or extremes of pH, the enzyme becomes **denatured** and the reaction stops. This is an irreversible change because the active site becomes distorted and no longer accepts the substrate.

Uses of Enzymes in the Home and Industry

Enzymes are cheap to use in industry and the home as they do not need high temperatures to work and can be reused.

Uses in the home include **biological washing powders** that contain enzymes produced by bacteria, e.g. proteases and lipases to digest protein and fat stains from clothes.

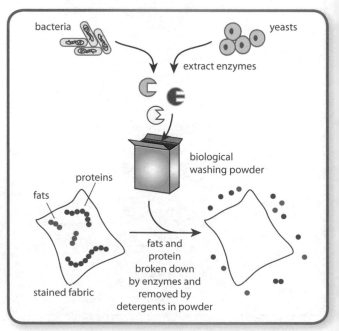

bacteria

yeasts

extract enzymes

biological washing powder

proteins

fats

stained fabric

fats and protein broken down by enzymes and removed by detergents in powder

In industry, enzymes help reactions to happen at low temperatures and pressures, helping to save energy and money for the manufacturer. However, most enzymes are denatured at high temperatures and some are costly to produce.

Examples of enzymes used in **industry** include:

● Carbohydrase enzymes used in the making of chocolate and syrup.

● Proteases used to pre-digest the protein in baby food.

● Isomerase (a carbohydrase) used to convert glucose syrup into the much sweeter fructose syrup used in much smaller quantities in slimming foods.

● Invertase (a sucrase enzyme) produced by yeast in the manufacture of sweets.

Digestion and Enzymes

Digestion is the breaking down of **large insoluble** molecules into **small soluble** molecules so that they can be absorbed into the bloodstream. This action is speeded up by **enzymes**. Enzymes are produced by specialised cells in **glands** and in the lining of the gut. They work outside the body cells.

Starch, proteins and fats are large, insoluble molecules.

● Starch is broken down into glucose by **carbohydrase** enzymes. Amylase (a carbohydrase) is produced in the salivary glands, the pancreas and the small intestine.

● Proteins are broken down into amino acids by **protease** enzymes (e.g. pepsin) produced in the stomach and the small intestine.

● Fats are broken down into fatty acids and glycerol by **lipase** enzymes. Lipase enzymes are produced by the pancreas and the small intestine.

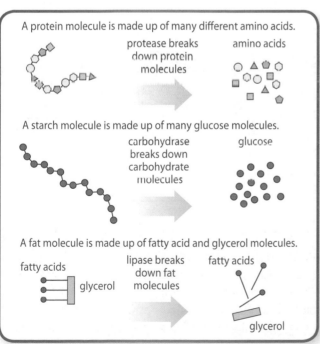

A protein molecule is made up of many different amino acids.

protease breaks down protein molecules — amino acids

A starch molecule is made up of many glucose molecules.

carbohydrase breaks down carbohydrate molecules — glucose

A fat molecule is made up of fatty acid and glycerol molecules.

fatty acids / glycerol — lipase breaks down fat molecules — fatty acids / glycerol

SUMMARY

● **Proteins have important roles in the body.**

● **Enzymes are biological catalysts.**

● **Enzymes are used in the body, at home and in industry.**

QUESTIONS

QUICK TEST

1. Name the building blocks of proteins.

2. Which type of enzyme breaks down fats?

3. Give an example of how proteases are used in industry.

EXAM PRACTICE

1. Explain why enzymes are used in industry.

   The quality of your written communication will be assessed in your answer to this question.
   (6 marks)

2. In the body, hydrogen peroxide is broken down into oxygen and water by an enzyme called catalase.

   Raw liver and fresh potato contain enzymes. In an experiment they were each added to test tubes of hydrogen peroxide and the number of bubbles of oxygen released were counted. The liver produced more oxygen bubbles in 10 minutes than the potato.

   a) Explain why it is important to use the same size cubes of potato and liver. **(2 marks)**

   b) Suggest one reason why the liver was more effective at breaking down the hydrogen peroxide. **(1 mark)**

Respiration and Exercise

Respiration supplies cells with a constant supply of energy so they can carry out cell processes and allow organs and systems to function.

Aerobic Respiration

Respiration is a series of chemical reactions in which the breakdown of glucose is used to release energy using oxygen. This takes place continuously in both plants and animals. Aerobic means 'with air' and as respiration needs oxygen we call it **aerobic respiration**.

The equation for respiration is:

> glucose + oxygen → carbon dioxide + water (+ energy)
>
> $C_6H_{12}O_6 + 6O_2 \rightarrow 6CO_2 + 6H_2O$

Energy may be used for growth, movement or to maintain body temperature in **endotherms** such as mammals and birds.

Anaerobic Respiration

During strenuous exercise, the oxygen supply is insufficient to meet the demands of the body, but energy is still needed. Respiration **without oxygen** is called **anaerobic respiration**. It produces **much less energy** and doesn't break down glucose completely. Some energy is lost as heat.

The word equation for anaerobic respiration is:

> glucose → lactic acid (+ energy)

Instead of carbon dioxide, **lactic acid** is produced. If muscles carry out vigorous activity, the lactic acid can contribute to muscle **fatigue**. An increased breathing rate supplies the body with enough oxygen to combine with the lactic acid and convert it to carbon dioxide and water. An increased heart rate ensures the blood carries lactic acid away to the liver where it is broken down when oxygen becomes available. The amount of oxygen needed to oxidise the lactic acid to carbon dioxide and water is called **excess post-exercise oxygen consumption** or **EPOC** (formerly known as **oxygen debt**).

Exercise and Fitness

People who exercise regularly are usually healthier than people who take little exercise.

Fitness is the ability to do physical activity, and **health** is being free from disease. Fitness can be measured in different ways: strength, stamina, flexibility, agility, speed and cardiovascular efficiency.

During exercise, breathing and pulse rates increase. The arteries supplying the muscles also dilate. This delivers oxygen and glucose more quickly to the respiring muscles and removes carbon dioxide via the blood.

Muscles store glucose as **glycogen**, which can then be converted back to **glucose** for use during exercise. Glucose can also be converted into fats that are stored under the skin and around organs as **adipose** tissue.

Exercise increases sweating and dehydration, which may lead to reduced sweating and a further increase in core body temperature.

A way to measure fitness could be to look at how long it takes for a person's pulse rate to return to normal after exercise, known as **recovery rate**. The quicker the recovery time, the fitter a person is.

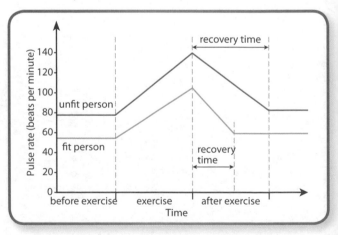

When starting an exercise programme, certain factors in a person's medical history or lifestyle should be taken into account. For example, symptoms, current medication, alcohol and tobacco consumption, level of physical activity, family medical history and previous treatments.

Any assessment of progress in a person's fitness needs to take into account the accuracy of the monitoring technique and the repeatability of the data obtained.

Breathing rate and heart rate monitors can be worn on the wrist or by the heart. They are more accurate at taking measurements than by hand because they produce a continuous measurement before, during and after exercise. There is also less chance of human error in the readings.

Injuries

Excessive exercise can cause sprains, dislocations and torn ligaments or tendons. A sprain happens when the ligament is overstretched or torn around the joint, for example, a twisted or sprained ankle.

'**RICE**' is used to treat sprains: Rest, Ice, Compression and Elevation. This helps to reduce pain and make recovery quicker.

A physiotherapist helps patients in the treatment of skeletal-muscular injuries to recover quickly and safely, and aims to prevent the situation from happening again.

Sports Drinks

Sports drinks contain sugars to replace the sugar used in energy release during the activity. They also contain water and ions to replace the water and ions lost during sweating.

If the water and ions are not replaced, the ion/water balance of the body is disturbed and the cells do not work as efficiently.

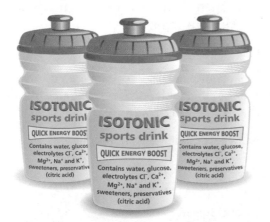

SUMMARY

- Aerobic respiration breaks down glucose using oxygen to release carbon dioxide, water and energy.
- Anaerobic respiration happens without oxygen.
- Breathing and pulse rates increase during exercise.

QUESTIONS

QUICK TEST

1. Write the word equation for aerobic respiration.

2. What are the products of anaerobic respiration?

3. Explain the term 'excess post-exercise oxygen consumption'.

EXAM PRACTICE

1. Look at the table. It shows the pulse rate in beats per minute of two athletes before exercise, immediately after exercise and then a further 5 minutes after exercise.

Athlete	Pulse rate (beats per minute)		
	Before	After	After 5 minutes
A	70	135	70
B	75	140	90

a) Which athlete is fitter? **(1 mark)**

b) How can you tell from the results? **(2 marks)**

c) What would be the most accurate way to measure pulse rate and why? **(2 marks)**

Movement

The Skeleton

The human skeleton is a living tissue that provides support, protection and allows movement to occur.

There are two types of skeleton – internal (**endoskeleton**) and external (**exoskeleton**). An internal skeleton begins life as cartilage and then ossification occurs where the cartilage is slowly replaced by the minerals calcium and phosphorus.

Whether a person is still growing or not can be determined by the amount of cartilage present. The internal skeleton is a rigid or semi-rigid structure that is moved by muscles and made up of bones and cartilage (ear, nose).

Humans have an internal skeleton that allows growth. Other animals, such as insects and spiders, have an exoskeleton made of a material called **chitin**. An exoskeleton only allows growth during a short period of time when the exoskeleton is shed during moulting.

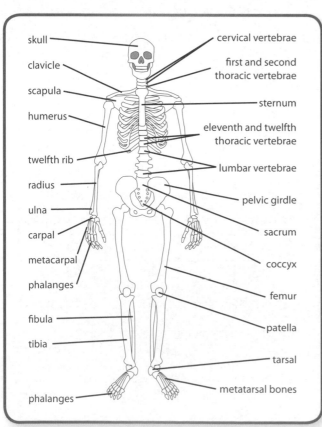

Some animals, such as worms, do not have a skeleton made of hard material, and some animals have an internal skeleton made only of cartilage, e.g. shark.

Bones

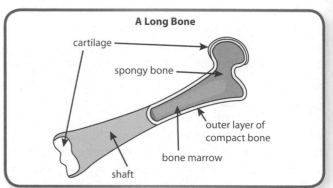

A long bone is longer than it is wide: it includes the femur, tibia and fibula of the leg and the radius, humerus and ulna of the arm. Long bones are hollow, which make them stronger than solid bones.

Despite being very strong, bones can be broken or fractured. A simple fracture occurs along one line splitting the bone into two pieces. Fractures can be **open** (**compound**) exposing the bone to contamination, or **closed** where the skin is intact.

A **greenstick fracture** can occur in children whose bones are still developing. Children's bones are not as brittle as adult bones and do not completely fracture but bend or bow in places.

X-rays are used to detect fractures. A bone fracture can also occur as a result of certain medical conditions that weaken the bones, such as **osteoporosis**. Elderly people are more prone to this.

Joints

Joints allow movement to occur when two bones meet. Bones are held together by **ligaments** and attached to muscles by **tendons**.

There are several different types of joint:

- a fixed joint – skull
- hinge joint – elbow, knee and wrist
- ball and socket joint – shoulder and hip.

Ball and socket and hinge joints are also known as **synovial joints**.

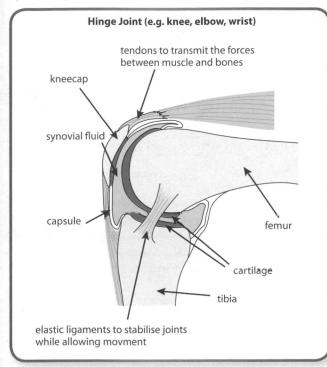

Hinge Joint (e.g. knee, elbow, wrist)

tendons to transmit the forces between muscle and bones

kneecap

synovial fluid

capsule

femur

cartilage

tibia

elastic ligaments to stabilise joints while allowing movment

The ends of these bones have a layer of smooth **cartilage** that acts as a shock absorber preventing the wearing away of the surfaces. The cartilage is covered by **synovial fluid** that helps reduce friction.

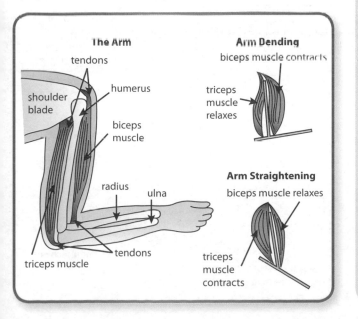

The Arm

tendons

humerus

shoulder blade

biceps muscle

radius   ulna

tendons

triceps muscle

Arm Bending

biceps muscle contracts

triceps muscle relaxes

Arm Straightening

biceps muscle relaxes

triceps muscle contracts

Muscles allow movement to occur at the joints. Muscles work in pairs called **antagonistic pairs**: when one contracts the other relaxes. An example is the biceps and triceps muscles that bend and straighten the arm.

SUMMARY

- **The skeleton provides support, protection and allows movement.**
- **Bones can be broken in three ways: open, closed and greenstick fractures.**
- **Joints allow movement where two bones meet.**
- **Muscles work in antagonistic pairs.**

QUESTIONS

QUICK TEST

1. What is the purpose of the cartilage and the synovial fluid in a joint?

2. Why is a long bone stronger than other bones?

3. Which animal has a skeleton made only of cartilage?

EXAM PRACTICE

1. Muscles allow movement to occur at joints.

   a) What type of joint is the shoulder joint? **(1 mark)**

   b) Describe what happens to the muscles in the arm when it bends. **(1 mark)**

   c) What do we call muscles that work in this way? **(1 mark)**

Central Nervous System

The nervous system allows humans to react to their surroundings and coordinate their behaviour.

The nervous system coordinates automatic processes, such as breathing and blinking.

The **central nervous system** (**CNS**) consists of the brain and spinal cord connected to different parts of the body by the **peripheral nervous system** (PNS) (**nerves**).

The body's sense organs contain specialised cells called **receptors**. Receptors detect changes in the environment called **stimuli**.

Receptors	Stimuli they detect
Receptors in the eyes	Sensitive to light
Receptors in the ears	Sensitive to sound and changes in position, which allows us to keep our balance
Receptors on the tongue and in the nose	Sensitive to chemicals, which allows us to taste and smell
Receptors in the skin	Sensitive to touch, pressure, pain and temperature changes

Light receptor cells, like most animal cells, have a nucleus, cytoplasm and cell membrane.

Receptors send **electrical impulses** along nerves to the brain and spinal cord in response to the stimuli from the environment. Nerve impulses travel in one direction only.

The CNS sends nerve impulses back along nerves to effectors which bring about a response. **Effectors** are muscles that result in movement, or they are glands that secrete hormones.

Nerves

Nerves are made up of nerve cells or **neurones**.

There are three types of neurone:

- **Sensory** neurones receive messages from receptors and send them to the CNS.

- **Motor** neurones send messages from the CNS to the effectors telling them what to do.
- **Relay** neurones connect sensory and motor neurones within the CNS.

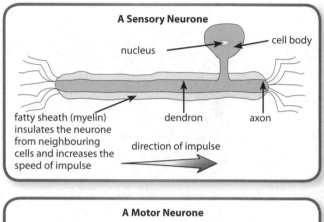

A Sensory Neurone

nucleus — cell body — dendron — axon — fatty sheath (myelin) insulates the neurone from neighbouring cells and increases the speed of impulse — direction of impulse

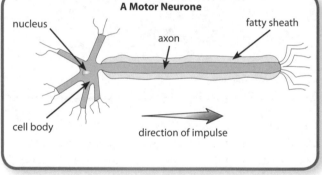

A Motor Neurone

nucleus — axon — fatty sheath — cell body — direction of impulse

Synapses

Between neurones there is a gap called a **synapse**.

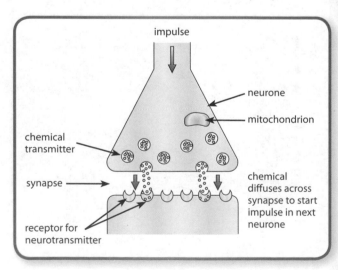

impulse — neurone — mitochondrion — chemical transmitter — synapse — receptor for neurotransmitter — chemical diffuses across synapse to start impulse in next neurone

When an impulse reaches the end of an axon, chemicals called **neurotransmitters** are released.

These chemicals diffuse across the gap and bind to specific receptor molecules on the membrane of the next neurone. This starts off an impulse in the next neurone.

Synapses can be affected by drugs (e.g. Ecstasy, beta blockers and Prozac) and alcohol, which slow down or prevent neurotransmitters from passing across.

The Reflex Arc

The reflex response to your CNS and back again can be shown in a diagram called the **reflex arc**.

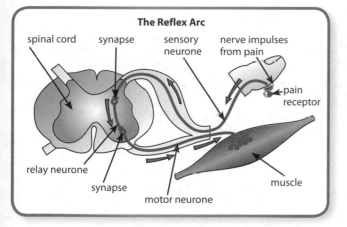

The Reflex Arc

spinal cord synapse sensory neurone nerve impulses from pain

pain receptor

relay neurone

synapse

motor neurone

muscle

The reflex arc can be shown in a block diagram:

stimulus → receptor → sensory neurone → relay neurone (in the spinal cord) → motor neurone → effector → response

In some circumstances the brain can override a reflex arc. For example, keeping hold of a hot object if you know that dropping it will hurt you more than the burn will.

Reflex and Voluntary Actions

Voluntary actions are things we have to think about – they are under conscious control and they have to be learned, like talking or writing.

Reflex actions produce rapid involuntary responses; they often protect us and other animals from harm. Examples include reflex actions in a newborn baby, the pupils' response to light, the knee jerk reflex and blinking. Simple reflex actions help animals survive as they respond to a stimulus such as smell to find food or to avoid predators.

SUMMARY

- The CNS consists of the brain and spinal cord.
- Receptors detect stimuli and effectors bring about a response.
- Motor, sensory and relay neurones transmit electrical impulses.
- Synapses are the junctions between neurones where neurotransmitters pass the impulse along.
- A reflex action is automatic and aids survival.

QUESTIONS

QUICK TEST

1. Name the **three** types of neurone.

2. Where would you find a synapse?

3. What is the stimulus that receptor cells in the eye detect?

EXAM PRACTICE

1. Fill in the gaps to show the path taken by a nerve impulse:

   stimulus → _____ → sensory

   neurone → relay neurone in the spinal cord →

   _____ neurone → _____

   → response   **(3 marks)**

2. Describe how the nerve impulse is transmitted from one neurone to another.   **(2 marks)**

3. Neurones are specialised cells. Describe myelin, its role in the neurone and its benefit. **(3 marks)**

The Eye

The eye is one of the human sense organs. Parts of the eye control the amount of light entering it and other parts control our focus on near and distant objects.

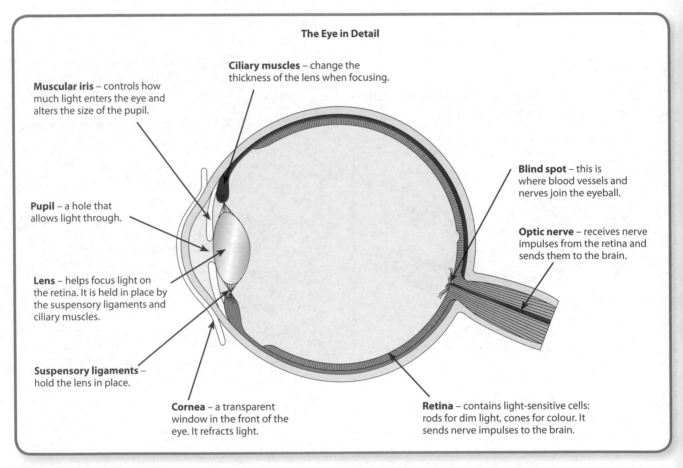

The Eye in Detail

Ciliary muscles – change the thickness of the lens when focusing.

Muscular iris – controls how much light enters the eye and alters the size of the pupil.

Blind spot – this is where blood vessels and nerves join the eyeball.

Optic nerve – receives nerve impulses from the retina and sends them to the brain.

Pupil – a hole that allows light through.

Lens – helps focus light on the retina. It is held in place by the suspensory ligaments and ciliary muscles.

Suspensory ligaments – hold the lens in place.

Cornea – a transparent window in the front of the eye. It refracts light.

Retina – contains light-sensitive cells: rods for dim light, cones for colour. It sends nerve impulses to the brain.

Adjusting to Light and Dark

Bright light

- Circular muscles contract. Radial muscles relax.
- The iris closes and makes the **pupil smaller**.

Dim light

- Radial muscles contract. Circular muscles relax.
- The iris opens and makes the **pupil bigger**.

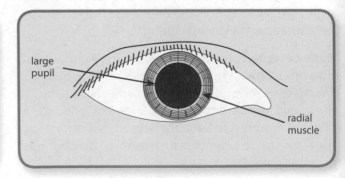

small pupil

circular muscle

large pupil

radial muscle

Focusing on Objects

Light passes through the cornea and lens and the object is focused on the retina. This is called **accommodation**.

	Ciliary muscles	Suspensory ligaments	Lens shape
Near objects	Contract	Slacken	Fat and round
Far objects	Relax	Contract	Thin and flat

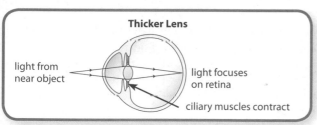

Thicker Lens

light from near object

light focuses on retina

ciliary muscles contract

Vision

Humans and many hunting animals have **binocular vision**. This means that our eyes are facing forward, which enables us to judge distances and depth effectively.

Cows, horses and other prey animals have **monocular vision** with eyes on the side of their heads. This allows them to have a wider field of view.

Problems with Vision

Short-sightedness results when the eyeball is too long. This means that light is focused too far in front of the retina. Sufferers can see near objects but not distant ones.

SUMMARY

- The pupil gets smaller in bright light and larger in dim light.
- The ciliary muscles and suspensory ligaments help us focus.
- Humans have binocular vision.
- People can have problems with their vision.

Long-sightedness is when the eyeball is too short and distant objects can be seen, but not close-up ones. Treatment for both short- and long-sightedness involves either…

 corneal surgery

using different lenses in contact lenses or glasses.

Red–green **colour blindness** is an inherited condition that affects more males than females. It is caused by the cones not functioning correctly.

Older people sometimes suffer from poor accommodation; they cannot focus quickly enough from close to distant objects because of weak ciliary muscles or stiff suspensory ligaments. This can cause problems judging distances when driving.

QUESTIONS

QUICK TEST

1. What is the difference between binocular vision and monocular vision?

2. Name the muscles that control the size of the lens.

3. What shape is the lens when focusing on near objects?

EXAM PRACTICE

1. Match each of the following parts of the eye with its function. **(2 marks)**

Iris	Changes shape to focus
Lens	Hold the lens in place
Suspensory ligaments	Controls how much light enters the eye

2. Describe how the eye accommodates. **(2 marks)**

3. Jean is an elderly lady who has difficulty with accommodation. What causes this difficulty and why might it be a problem? **(3 marks)**

The Brain

The brain is situated at the top of the spinal cord and is protected by the skull.

The brain and spinal cord make up the **central nervous system**.

The evolution of the larger brain gave early humans a better chance of survival. The **cerebral cortex** makes up the outer layer of the brain. It is the part of the brain most concerned with intelligence, memory, language and consciousness.

The **medulla oblongata** is the part of the brain that attaches to the spinal cord. It controls automatic actions such as breathing and heart rate. The **cerebellum** controls our coordination and balance.

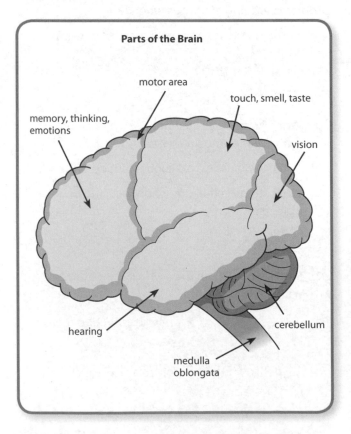

Parts of the Brain

- motor area
- touch, smell, taste
- memory, thinking, emotions
- vision
- hearing
- cerebellum
- medulla oblongata

Scientists can map regions of the brain to particular functions. This has been achieved by studying patients with brain damage, stimulating different parts of the brain with electricity and by using brain scans such as **MRI** that show brain structure and activity.

Learning

The brain works by sending electrical impulses received from the sense organs to the effectors (muscles and glands). It coordinates the response.

Mammals have a complex brain of billions of neurones that allows learning by experience and behaviour through interaction with the environment. This results in nerve pathways forming in the brain. When we learn from experience, some pathways in the brain become more likely to transmit impulses than others, so it is easier to learn through repetition.

The variety of potential pathways makes it possible for animals to adapt to new situations.

The way in which humans learn language has long been debated by linguists and child psychologists. Some say that there is a crucial period of language acquisition that ends when a child is around 12 years of age.

Memory

Memory is the storage and retrieval of information. Memory can be divided into either short-term or long-term memory.

Humans are more likely to remember information if…

- they can see a pattern
- there is repetition
- there is a strong stimulus associated with the information (e.g. colour, light, smell or sound).

Models can be used to describe memory. For example, Atkinson and Shiffrin in the 1960s presented the **Multistore Model of Memory**. They suggested that memory was made up of three stores: the sensory memory store, the short-term memory and the long-term memory. Each store was described to have a specific function: information is received through the sensory store, then rehearsed in the short-term memory and transferred to long-term memory.

Scientists have produced lots of models to try and explain how memory works. Models are limited and so far none have provided an exact explanation.

The Multistore Model of Memory

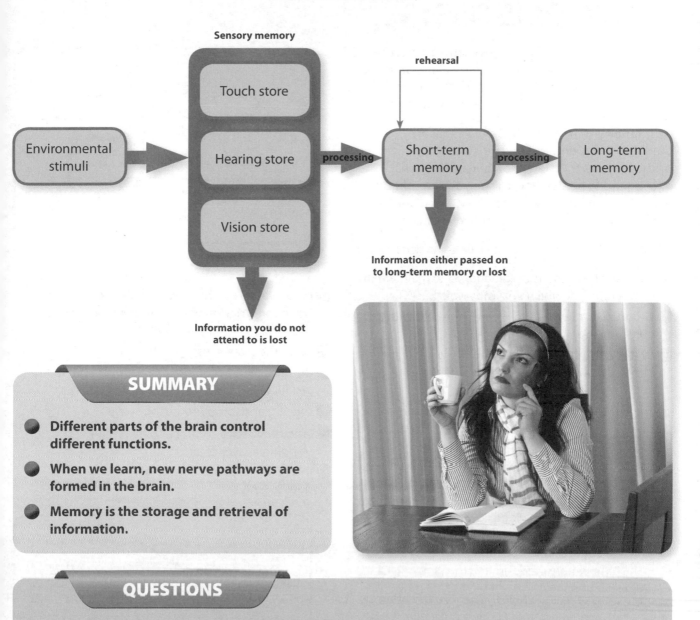

Sensory memory

Touch store

Environmental stimuli

Hearing store

processing

Vision store

rehearsal

Short-term memory

processing

Long-term memory

Information either passed on to long-term memory or lost

Information you do not attend to is lost

SUMMARY

- **Different parts of the brain control different functions.**

- **When we learn, new nerve pathways are formed in the brain.**

- **Memory is the storage and retrieval of information.**

QUESTIONS

QUICK TEST

1. Which part of the brain controls breathing and heart rate?

2. Which part of the brain controls coordination and balance?

3. Name the three stores that are used in the Multistore Model of Memory.

EXAM PRACTICE

1. We learn through experience and behaviour. How does repeating an experience help us learn? **(1 mark)**

2. Why is the large variety of potential pathways an advantage? **(1 mark)**

Homeostasis

The nervous system and hormones enable us to respond to **external** changes in the environment and monitor and change our **internal** environment so that conditions stay at safe, stable levels. This is called **homeostasis**.

Hormones are chemical messengers that are produced in the endocrine glands and are transported by the blood to their target organism. Hormonal effects tend to be slow, long lasting and can affect a number of organs. Nervous control is much quicker, doesn't last very long and is confined to one area.

Internal conditions of the body that are controlled include:

- ⚫ **Temperature** (thermoregulation) – to maintain the temperature at which enzymes work best.
- ⚫ **Blood sugar levels** – to provide the cells with a constant supply of energy.
- ⚫ **Water content** (osmoregulation) – water leaves the body when we breathe out and when we sweat; excess water is lost in urine.
- ⚫ **Ion content** – ions are lost in our sweat and urine.

Thermoregulation

Warm-blooded animals have mechanisms that can keep body temperature constant. The **hypothalamus** in the brain (thermoregulatory centre) has receptors sensitive to the temperature of the blood flowing through it. The **skin** has receptors that send impulses to the brain giving information about skin temperature. Your body aims to keep the temperature inside around the same, **37 °C**.

If body temperature is too high this can cause heat stroke, dehydration and even death. If body temperature is too low this can cause **hypothermia** and death if untreated.

When it's hot:

- ⚫ blood vessels at the surface of the skin **widen** (**vasodilation**) allowing more blood to flow to the surface
- ⚫ heat **radiates** from the skin and is lost
- ⚫ sweat glands secrete sweat, which **evaporates** from the skin and takes away **heat energy**
- ⚫ hair erector muscles relax, so hairs lie flat allowing heat to escape.

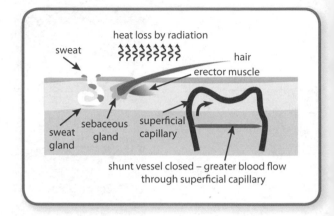

shunt vessel closed – greater blood flow through superficial capillary

When it's cold:

- ⚫ blood vessels at the surface of the skin **contract** (**vasoconstriction**) so very little blood reaches the surface
- ⚫ very little heat is lost by radiation
- ⚫ muscles contract quickly (shivering), which produces extra heat
- ⚫ sweat glands stop producing sweat
- ⚫ erector muscles contract, so hairs stand up trapping a layer of air
- ⚫ increased respiration helps generate heat, as does exercise.

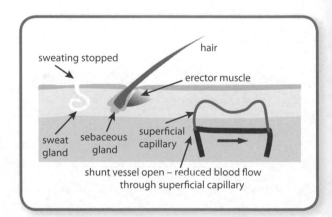

shunt vessel open – reduced blood flow through superficial capillary

Many warm-blooded animals have a thick layer of fat beneath their skin for insulation.

Some effectors work antagonistically (in opposite pairs), which allows a more sensitive and controlled response.

Blood Sugar Level and Diabetes

Glucose (sugar) is needed for respiration. The **pancreas** regulates the level of glucose in the blood by secreting two hormones: **insulin** and **glucagon**. High levels of sugar, common in some processed foods, are quickly absorbed into the bloodstream, causing a rapid rise in blood sugar level.

The **liver** responds to insulin by taking up excess glucose and storing it as **insoluble glycogen**.

Glucagon stimulates the conversion of stored glycogen in the liver back into glucose when the blood sugar level falls too low. This is an example of **negative feedback**.

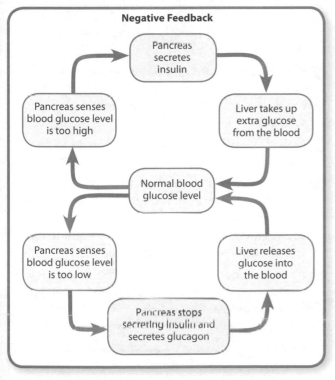

Negative Feedback

Pancreas secretes insulin

Pancreas senses blood glucose level is too high

Liver takes up extra glucose from the blood

Normal blood glucose level

Pancreas senses blood glucose level is too low

Liver releases glucose into the blood

Pancreas stops secreting insulin and secretes glucagon

Type 1 diabetes is a condition in which a person's blood glucose level may rise to fatally high levels because the pancreas doesn't make enough insulin. This can be controlled by attention to diet, regular exercise and/or insulin injections. Type 1 diabetes is diagnosed by the presence of glucose in urine.

Type 2 diabetes is caused by a person becoming resistant to insulin. This can be controlled by attention to diet (high in fibre and complex carbohydrates) and physical activity. There is evidence to suggest a correlation between obesity and type 2 diabetes.

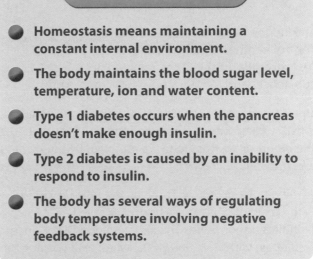

SUMMARY

- Homeostasis means maintaining a constant internal environment.

- The body maintains the blood sugar level, temperature, ion and water content.

- Type 1 diabetes occurs when the pancreas doesn't make enough insulin.

- Type 2 diabetes is caused by an inability to respond to insulin.

- The body has several ways of regulating body temperature involving negative feedback systems.

QUESTIONS

QUICK TEST

1. Where is insulin made?

2. What effect does glucagon have on the blood sugar level?

3. What effect does sweating have?

EXAM PRACTICE

1. Tom was walking outside in the cold; he looked pale and was shivering.

    a) Give two ways in which his body was working to try and keep him warm. **(2 marks)**

    b) Which part of the brain monitors body temperature? **(1 mark)**

    c) At which temperature is our body maintained? Choose the correct answer:
    35°C   37°C   39°C   42°C   **(1 mark)**

2. A healthy person, who does not have diabetes, exercises for an hour. Their blood sugar level decreases below the normal level. Explain how blood sugar level returns to normal.

    The quality of your written communication will be assessed in your answer to this question.

    (6 marks)

The Kidney

The kidney has a major role in homeostasis by controlling the amount of water in our body (**osmoregulation**) and the removal of excess substances and the poisonous urea (**excretion**).

Urea is produced in the liver from excess amino acids. Ions are taken into the body in food and absorbed into the blood; any excess ions are removed by the kidneys and in sweat.

Water regulation is monitored by the **pituitary gland** in the brain that releases a hormone called ADH (antidiuretic hormone). The kidneys and ADH balance water gain with water loss. If the water or ion content of the body is wrong, too much water may move into or out of cells and damage them.

Ultrafiltration

- The blood arrives in the **renal artery** at **high pressure** and enters the group of capillaries called the **glomerulus**.
- High pressure squeezes water, urea, ions and glucose out of the blood into the **Bowman's capsule**. Large molecules stay in the blood.
- Liquid in the first **tubule** contains useful substances like glucose and some ions. These must be **reabsorbed**.
- As the blood enters the first tubule, useful substances are reabsorbed by **active transport**. The cleaned blood enters the renal vein and leaves the kidney.
- All urea, excess ions and water pass into the collecting duct. This fluid continues out of the tubule into the ureter and down to the bladder as **urine**.
- **ADH** is released when more water needs to be reabsorbed; the result is more concentrated urine.
- When ADH production stops, there is excess water in the blood. The result is dilute urine. This is an example of **negative feedback**. Excess water is also removed in sweat, faeces and breathing out. This affects how much urine is produced.
- When it is hot or during exercise more water is lost through sweat, so more water has to be taken in to balance this loss.

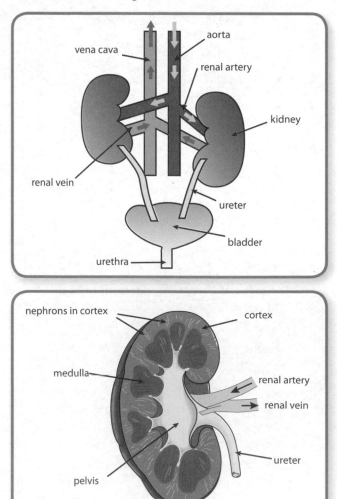

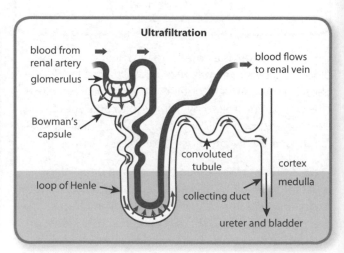

Kidney Failure

The presence of blood cells in urine may indicate disease in the kidney. There are two treatments for kidney failure – a **transplant** or **dialysis**.

A transplant enables a diseased kidney to be replaced with a healthy one from a donor. There is less chance of the immune system rejecting the kidney if the tissue type from the donor is a close match to the recipient and the recipient takes anti-rejection drugs that suppress the immune system.

During dialysis, a machine is used that filters the patient's blood and restores the concentration of dissolved substances in the blood to normal levels. The blood flows between partially permeable membranes separating the blood and dialysis fluid in the machine.

Dialysis fluid contains the same concentrations of substances as the blood so that useful substances are not lost. Urea passes out of the patient's blood into the fluid.

	Advantages	Disadvantages
Dialysis	Blood is 'cleaned' No surgery involved	Difficult to travel, e.g. foreign holidays Transport to and from dialysis Time consuming – must take place regularly
Transplant	Freedom from dialysis Diet is less limited Better quality of life (able to travel) Feel better physically	Possible rejection Immune system suppressed Risks of major surgery (anaesthetic, infection, etc)

SUMMARY

- The kidney filters the blood, removes waste and regulates water levels.
- Antidiuretic hormone (ADH) released by the pituitary gland regulates water levels.
- Kidney failure can be treated by dialysis or transplant.

QUESTIONS

QUICK TEST

1. What is osmoregulation?

2. What treatments are available for kidney failure?

3. Where is ADH produced?

EXAM PRACTICE

1. Urea is a toxic substance that needs to be removed from the body in urine.

   a) Where is urea formed in the body and from what substance? **(2 marks)**

   b) Describe how the kidneys remove urea from the body. **(4 marks)**

   c) List two other substances found in a healthy person's urine. **(2 marks)**

2. Shaun goes for a long bike ride on a very hot day. He forgets to take a drink with him.

   a) What would you expect his urine to be like that evening? **(2 marks)**

   b) Describe the role of ADH in preventing Shaun from losing too much water. **(3 marks)**

Breathing and the Lungs

Small, simple organisms such as amoebas and earthworms take in oxygen through their moist permeable external surfaces.

Larger, more complex animals have specialised organs such as **gills** and **lungs**.

In fish, water flows in the opposite direction to the blood capillaries in the gills. The concentration of oxygen is lower in the blood than the water so this movement allows oxygen to diffuse into the blood. This is called the **counter current** exchange system and is helped by the large surface area of the gills.

The Lungs

The lungs are situated in the upper part of the body called the **thorax**. The thorax is separated from the lower part called the **abdomen** by the diaphragm and protected by the ribcage.

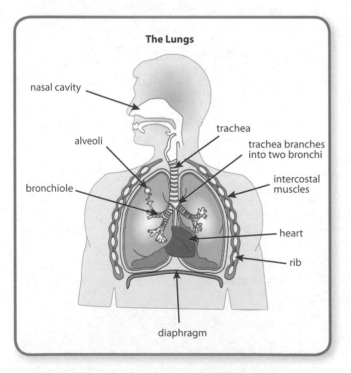

The Lungs

- nasal cavity
- alveoli
- bronchiole
- trachea
- trachea branches into two bronchi
- intercostal muscles
- heart
- rib
- diaphragm

Our cells produce carbon dioxide that is transported to the lungs by the blood. The lungs remove the toxic carbon dioxide by diffusion from the blood into the alveoli and then we breathe it out. Oxygen diffuses in the opposite direction. This is called **gas exchange**.

The alveoli in the lungs are well designed for their job of gas exchange:

- There are millions of them, which present a **large surface area**.
- They are in **very close contact** with lots of blood capillaries, which maintains a concentration gradient.
- Their surface lining is **moist** so the gases can dissolve before they diffuse across the **thin membrane**.

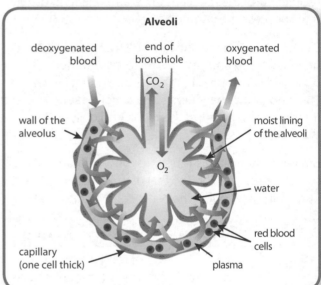

Alveoli

- deoxygenated blood
- end of bronchiole
- oxygenated blood
- CO_2
- wall of the alveolus
- moist lining of the alveoli
- O_2
- water
- red blood cells
- capillary (one cell thick)
- plasma

Ventilation

Breathing in and out is called **ventilation**. Two sets of intercostal muscles between the ribs and the diaphragm (which is also a muscle) help us breathe.

Inspiration is breathing in and **expiration** is breathing out.

	Ribs	Diaphragm	Volume of chest cavity
Breathing in	Move up and out	Contracts and moves down	Increases
Breathing out	Move down and in	Relaxes and moves up	Decreases

Lime water is often used in experiments to show the presence of carbon dioxide. It turns from clear and colourless to cloudy when carbon dioxide is present.

A **spirometer** is used to measure lung capacity. **Tidal volume** is the amount of air breathed in or out of the lungs in one breath.

Vital capacity is the maximum amount of air that can be forcibly exhaled after breathing in. **Residual volume** is the amount of air remaining in the lungs after exhaling. **Total lung capacity** is the maximum volume of air in the lungs.

Composition of Gases

Gas	Inhaled air %	Exhaled air %
Oxygen	21	16
Carbon dioxide	0.04	4
Nitrogen	79	79
Water vapour	Varies	High

Lung Disease

The respiratory system protects itself from disease by producing **mucus** and having **ciliated cells** lining the trachea and bronchi. The mucus traps dust and some microorganisms and the ciliated cells move it up away from the lungs.

Smoking prevents cilia from working properly so mucus is not cleared from the respiratory tract. This can lead to a 'smoker's cough'.

The respiratory system is prone to disease as the lungs are a 'dead end'. Diseases of the lungs include pneumonia, bronchitis, lung cancer, asbestosis, cystic fibrosis and asthma.

Asbestosis is an asbestos-related lung disease causing inflammation and scarring in the airways as well as increasing the risk of certain cancers.

Asthma is caused by the airways constricting and becoming inflamed. There is also an excess of mucus. This causes wheezing, shortness of breath, chest tightening and coughing.

Bronchodilators (inhalers) work by opening up the constricted airways. Allergens, cold or warm air, viral illness, exercise or stress can trigger an asthma attack.

SUMMARY

- The lungs are adapted for gas exchange.
- The alveoli are efficient gas exchange surfaces.
- The lungs are prone to many diseases.

QUESTIONS

QUICK TEST

1. What is ventilation?

2. Explain the term 'gas exchange'.

3. Which muscles are involved in ventilation?

EXAM PRACTICE

1. Mandy suffers from asthma.

   a) Why are the lungs prone to diseases such as asthma? **(1 mark)**

   b) List three symptoms of an asthma attack. **(3 marks)**

   c) How does the respiratory system protect itself from disease? **(3 marks)**

2. Describe how the alveoli are efficient at gas exchange. **(3 marks)**

Controlling Fertility

The menstrual cycle lasts approximately 28 days. It consists of a menstrual bleed and ovulation (the release of an egg).

The Menstrual Cycle Stages

Hormones control the whole cycle. Ovaries secrete the hormones **progesterone** and **oestrogen**.

- **Days 1–5**: a menstrual bleed (a period) occurs. The lining of the uterus breaks down, caused by a lack of oestrogen and progesterone.
- **Days 5–14**: oestrogen is released and the uterus lining builds up again. Oestrogen stimulates egg development and release of the egg from the ovaries – this is called **ovulation**.
- **Days 14–28**: progesterone is released which maintains the uterus lining. If no fertilisation occurs, progesterone production stops.
- **Days 28–5**: the cycle begins again.

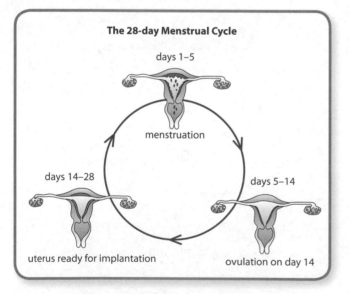

The 28-day Menstrual Cycle

days 1–5

menstruation

days 14–28

days 5–14

uterus ready for implantation

ovulation on day 14

The Pituitary Gland

The hormones released from the ovaries are controlled by the **pituitary gland**, which is situated at the base of the brain.

The pituitary gland secretes two hormones: **follicle stimulating hormone (FSH)** and **luteinising hormone (LH)**.

FSH stimulates follicles to become mature and triggers the production of oestrogen from the ovaries. Oestrogen is responsible for the repair of the uterus wall in preparation for a fertilised egg.

The high levels of oestrogen signal the release of LH. LH causes ovulation (the release of an egg on day 14) and the release of progesterone by the corpus luteum.

Progesterone maintains the lining of the uterus and inhibits FSH and LH production. During pregnancy, levels of progesterone remain high. If no pregnancy occurs, progesterone production stops and the cycle begins again.

Menstruation is triggered by a drop in oestrogen and progesterone levels. Low levels of progesterone allow an increase in FSH. The menstrual cycle is an example of **negative feedback**.

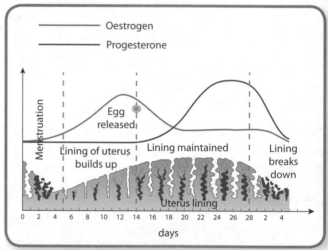

— Oestrogen
— Progesterone

Menstruation

Egg released

Lining of uterus builds up

Lining maintained

Lining breaks down

Uterus lining

0 2 4 6 8 10 12 14 16 18 20 22 24 26 28 2 4

days

Controlling Fertility

Fertility in women can be controlled using artificial hormones.

Oral contraceptives may contain oestrogen and progesterone to inhibit FSH so that no eggs mature. The first birth control pills contained large amounts of oestrogen, which resulted in women suffering side effects. Today, they contain much lower doses of oestrogen, or are progesterone only as this has fewer side effects.

Fertility Treatments

FSH and LH can be administered as a 'fertility drug' to women whose own production is too low to stimulate eggs to mature. Sometimes the use of FSH can result in multiple births.

In vitro **fertilisation** (**IVF**) is a treatment for infertile couples. FSH and LH are given to the woman to stimulate the maturation of several eggs.

The eggs and sperm are collected and fertilisation occurs outside the body. Cells that develop are then implanted in the uterus for growth and development into an embryo.

IVF is an expensive process and is not always successful. There are some social and ethical implications of IVF, particularly its use with older couples.

Some clinics set an age limit. They feel the chances of success fall with increasing age, and also the possibility of birth defects increases with older eggs.

IVF can create many fertilised embryos. This raises the question of what to do with the ones that are not implanted in the womb.

Some couples look at freezing them for later use, but what about the rest? Should they be thrown away or used for medical research? What if a couple splits up? Who owns the embryos?

Other treatments for infertility include egg donation, artificial insemination, surrogacy and ovary transplants.

SUMMARY

- The menstrual cycle is controlled by hormones.
- Fertility can be controlled using artificial hormones.
- Fertility treatment, including IVF, can help women become pregnant.

Antenatal Testing

Foetal development can be checked to identify conditions such as Down's syndrome using **amniocentesis** and chromosomal analysis. Foetal screening raises ethical issues.

QUESTIONS

QUICK TEST

1. Where are the hormones progesterone and oestrogen made?

2. Name the two female sex hormones produced by the pituitary gland.

3. What is ovulation?

EXAM PRACTICE

1. Fill in the gaps using the words below.
   (4 marks)

   oestrogen   progesterone   FSH   LH

   _____ begins the menstrual cycle and causes the ovaries to mature an egg and release _____ . This in turn inhibits FSH but causes the release of _____ . This hormone causes ovulation and the release of _____ that maintains the uterus lining.

2. Oral contraceptives prevent the ovaries from producing eggs.

   a) Which hormones do they contain?
      (1 mark)

   b) Which hormone do they prevent from being produced? **(1 mark)**

   c) Where is this hormone produced?
      (1 mark)

Pathogens

Microorganisms include bacteria, viruses, fungi and single-celled organisms called protozoa.

Most microorganisms are harmless and many perform vital functions. However, some microorganisms, called **pathogens**, cause infectious disease.

Pathogens are spread through the following:

- Infected **food and water** – e.g. cholera from infected drinking water and *Salmonella* bacteria that causes food poisoning.
- The **air** (e.g. through sneezing) – e.g. flu virus, colds and pneumonia.
- **Contact** with infected people or objects used by infected people – e.g. athlete's foot fungus.
- **Bodily fluids**, e.g. HIV.
- **Animal vectors**, e.g. dysentery bacterium from houseflies.

Non-infectious diseases are not caused by pathogens but other causes, such as genetics.

Bacteria

Bacteria are living organisms that feed, move and carry out respiration. They reproduce rapidly in favourable conditions, producing very large numbers that are exact copies of themselves.

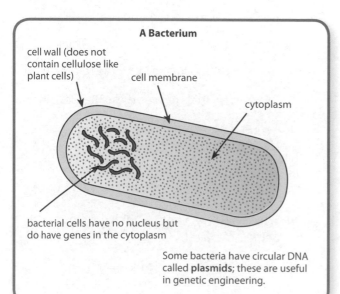

A Bacterium

cell wall (does not contain cellulose like plant cells)

cell membrane

cytoplasm

bacterial cells have no nucleus but do have genes in the cytoplasm

Some bacteria have circular DNA called **plasmids**; these are useful in genetic engineering.

Bacteria do not contain a nucleus. Their genetic information (DNA) is free within the cytoplasm as either chromosomes or **plasmids**. Some bacteria have flagella to propel them. They do not contain mitochondria or chloroplasts.

How Bacteria Cause Disease

Bacteria **destroy living tissue**. For example, tuberculosis destroys lung tissue. Bacteria can produce poisons called **toxins**. Food poisoning is an example.

Viruses

Viruses consist of a protein coat surrounding a few genes.

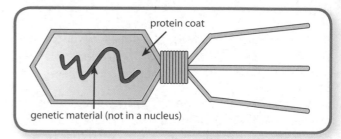

protein coat

genetic material (not in a nucleus)

Viruses are much smaller than bacteria. They do not feed, move, respire or grow: they just reproduce.

Viruses can only survive inside the cells of a living organism. They reproduce inside the host cell and release thousands of new viruses to infect new cells, killing the cell in the process. Examples of diseases caused by viruses include HIV, flu, chicken pox and measles.

Fungi

Fungi cause athlete's foot and ringworm. They reproduce by making **spores** that can be carried from person to person.

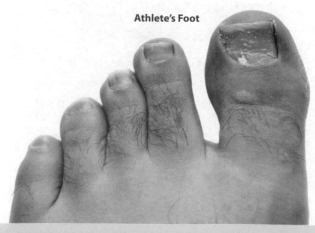

Athlete's Foot

Protozoa

Protozoa are tiny and made up of only one cell. Food contaminated with protozoa may cause dysentery. Malaria is a disease caused by protozoa living in the blood.

How Do Pathogens Get In?

Pathogens have to enter our body before they can do any harm. Pathogens can enter through the:

- skin – when damaged
- digestive system – via food and drink
- respiratory system – breathed in
- reproductive system – through sexual intercourse.

Vectors

Some pathogens rely on **vectors** to transfer them from one organism to another, e.g. *Anopheles* **mosquitoes**.

The mosquito feeds on the blood of an infected person and sucks in a protozoan called *Plasmodium*. When the mosquito bites another person it injects the *Plasmodium*, passing on the infection. This is an example of protozoa being a parasite.

A **parasite** is an organism that grows and feeds off another without any benefit to the host. Examples include fleas, head lice, tapeworms and mistletoe.

Knowing the life cycle of a disease and how it spreads can help control infections.

Cancer

Cancer is a condition caused when body cells grow out of control and become a mass of cells known as a **tumour**. Tumours that stop growing are **benign**; ones that invade the surrounding tissues and organs are called **malignant**.

Changes in diet and lifestyle may reduce the risk of some cancers.

SUMMARY

- Some microorganisms, called pathogens, can cause infectious diseases.
- Pathogens can enter the body through damaged skin, and the digestive, respiratory or reproductive system.
- Parasites are organisms that grow and feed off another organism without any benefit to the host.

QUESTIONS

QUICK TEST

1. Name four types of microorganism.
2. What is a vector? Give an example.
3. What do we call microorganisms that cause disease?

EXAM PRACTICE

1. Malaria is an infectious disease.

   a) Which type of microorganism is malaria caused by? **(1 mark)**

   b) Describe how malaria is transmitted. **(4 marks)**

   c) Suggest how the spread of malaria can be controlled. **(2 marks)**

2. Which of the following is caused by fungi? **(1 mark)**

   A Athlete's foot   B Flu
   C Cholera   D Anaemia

3. Which is an example of an infectious disease? **(1 mark)**

   A Diabetes   B Cancer
   C Tuberculosis   D Kwashiorkor

Natural Defences

The human body has many methods of **preventing pathogens** from entering the body, including physical barriers and chemical defences.

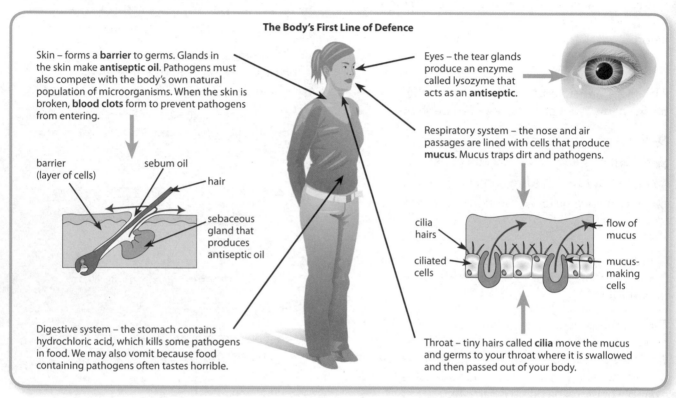

The Body's First Line of Defence

Skin – forms a **barrier** to germs. Glands in the skin make **antiseptic oil**. Pathogens must also compete with the body's own natural population of microorganisms. When the skin is broken, **blood clots** form to prevent pathogens from entering.

barrier (layer of cells)

sebum oil

hair

sebaceous gland that produces antiseptic oil

Digestive system – the stomach contains hydrochloric acid, which kills some pathogens in food. We may also vomit because food containing pathogens often tastes horrible.

Eyes – the tear glands produce an enzyme called lysozyme that acts as an **antiseptic**.

Respiratory system – the nose and air passages are lined with cells that produce **mucus**. Mucus traps dirt and pathogens.

cilia hairs

ciliated cells

flow of mucus

mucus-making cells

Throat – tiny hairs called **cilia** move the mucus and germs to your throat where it is swallowed and then passed out of your body.

Preventing the Spread of Germs

We can prevent the spread of germs by:

- **sterilising** equipment used in food preparation or operating theatres, by heating them to 120°C
- **disinfecting** work surfaces and areas like toilets
- using **antiseptics** which can kill pathogens if we cut ourselves
- having general good **hygiene**, as this is important in preventing the spread of disease.

In 1847, Dr. **Semmelweiss** suggested that washing hands prevented infection. He noticed that the number of deaths from infectious diseases in his hospitals significantly reduced when doctors washed their hands between treating patients. He suggested the modern-day idea of disinfecting hands and instruments.

Many scientists were involved in pioneering work in the treatment of disease. Louis **Pasteur** discovered that germs cause disease, Joesph **Lister** developed the use of antiseptics and Alexander **Fleming** discovered penicillin, an antibiotic.

Some medicines, including painkillers, help to relieve the symptoms of infectious disease, but do not kill the pathogens. Some plants produce chemicals that have antibacterial effects in order to defend themselves, some of which are used by humans, e.g. tea tree oil.

Natural Disasters and Disease

Natural disasters, such as volcanic eruptions, tsunamis or earthquakes, can cause a rapid spread of disease. This can happen if there has been damage to sewage systems and water supplies, displacement of people, disruption to health services, or damage to electrical supplies causing rapid food decay.

The Immune System Response

If pathogens do get into the body, **white blood cells** travelling in the blood are activated:

- White blood cells make chemicals called **antitoxins** that destroy the **toxins** produced by bacteria and prevent inflammation.
- White blood cells called **phagocytes** engulf some bacteria or viruses before they have a chance to do any harm. This is the body's **second line of defence** and is **non-specific**.
- If large numbers of pathogens enter, then a type of white blood cell called **lymphocytes** become involved. This is the **third line of defence** and is called the **specific immune response**.

All pathogens have proteins on their surface called **antigens**. Foreign antigens trigger a response by some white blood cells (lymphocytes) that secrete **antibodies** specific to the antigen. They destroy them in various ways, like clumping them together so that phagocytes can engulf and destroy them. Antibodies are made specifically to fight a particular antigen.

Natural Immunity

Making antibodies takes time. Initially, an infected person feels ill, and then gets better because the disease is destroyed by the white blood cells. Once a particular antibody is made, a few stay in the body and act as a memory. If the same disease-causing organism enters the body again, the antibodies are much quicker at multiplying and destroying it and the person may feel no symptoms. The person is then **immune** to that disease.

Monoclonal Antibodies and Hybridomas

In response to different antigens, white blood cells produce different antibodies. Each antigen has several areas on its surface, so one antigen may require several different types of antibodies (**polyclonal antibodies**).

Monoclonal antibodies bind only to a particular antigen or part of an antigen. The white blood cells that produce monoclonal antibodies cannot divide. Instead they fuse with a cancerous cell called a **myeloma** that can divide. The fused cell is called a **hybridoma**, which produces monoclonal antibodies and divides.

Monoclonal antibodies are used in **pregnancy testing**, in diagnosis and locating blood clots and cancer cells, and in the treatment of some cancers. Monoclonal antibodies are useful because they target specific cells compared to other drug and radiotherapy treatments.

SUMMARY

- The human body has several ways of preventing pathogens from entering the body.
- The immune system, involving white blood cells, responds to pathogens that enter the blood.
- Natural immunity occurs when a person develops immunity after suffering from a disease.

QUESTIONS

QUICK TEST

1. What chemicals do white blood cells produce to fight toxins?

2. How does the body recognise foreign bacteria and viruses?

3. Name four ways of preventing the spread of pathogens.

EXAM PRACTICE

1. Rachel has cut her finger and it has become infected with bacteria. Describe how phagocytes can help get rid of the infection.
   (1 mark)

2. Distinguish between the following:

   a) Antigens and antibodies **(2 marks)**

   b) Phagocytes and lymphocytes **(2 marks)**

3. What is meant by the term 'specific immune response'?
   (1 mark)

Artificial Immunity

There are various ways of treating disease once the pathogens get past the body's natural defences. There are also ways to prevent disease.

Vaccination

Vaccination, or **immunisation**, was discovered in 1796 by Edward **Jenner**, who found that infecting people with cowpox protected them from getting smallpox. It would be difficult to carry out Jenner's work today because of the ethics involved.

Vaccines contain dead or harmless pathogens that still have antigens (or parts of antigens), and they are injected into the body. White blood cells respond to the antigens by multiplying and producing antibodies. The antigens also trigger the production of memory lymphocytes. Vaccines may be produced that protect against bacteria and viruses.

A vaccine is like an advance warning, so that if the person is infected by the pathogen again, the white blood cells can respond immediately.

Vaccinations are **passive immunity** as the body produces its own antibodies in response to the vaccine. An injection of ready-made antibodies is called **active immunity**.

Vaccines help prevent the spread of disease if a large proportion of the population is immunised, but people have the right to choose whether to be vaccinated or not. It is important for people to make decisions based on sound scientific evidence rather than the effect of media and public opinion.

During the 1990s, the viral vaccine used to treat **MMR** (measles, mumps and rubella) caused controversy. There was concern about the side effects of using a triple vaccine instead of three separate ones. This was later disproved. Vaccines can never be risk free, so the benefits of the vaccine have to be weighed against its risks.

Antimicrobials

Antimicrobials is a general term describing chemicals that kill, or inhibit, bacteria, fungi and viruses.

Sometimes bacteria breach the body's defences and reproduce, making you feel ill. **Antibiotics** help cure bacterial diseases by killing the infecting bacteria or preventing their growth inside your body.

Penicillin was the first form of antibiotic. It is made from a **fungi** called *Penicillium notatum*. Antibiotics can only treat bacterial infections; they cannot treat viral infections. The body usually has to fight viruses on its own, but scientists are slowly developing **antiviral** drugs.

Antifungals can be used to treat fungal infections such as athlete's foot.

Antibiotic Resistance

It is important that specific bacteria are treated by specific antibiotics. The use of antibiotics has greatly reduced deaths from infectious bacterial disease. Inappropriate and overuse of antibiotics, however, has increased the development of antibiotic **resistant** strains of bacteria.

Strains of bacteria, including **MRSA**, have developed resistance to antibiotics as a result of natural selection. **Mutations** of pathogens produce new strains that are no longer affected by the antibiotic.

Antibiotics kill individual pathogens of the non-resistant strain. Individual resistant pathogens survive and reproduce, so the population of resistant strain increases. The new strain spreads rapidly because people are not immune to it and there is no effective treatment. New antibiotics are constantly needed.

Nowadays, antibiotics are not used to treat non-serious infections in order to slow down the rate of development of resistant strains. Places such as hospitals need scrupulous hygiene to try and prevent resistant strains developing and spreading, and they need to avoid the overuse of antibiotics.

Drug Testing

Scientists continually develop new drugs. People will react differently to drugs and vaccines because of genetic differences. Drugs have to be extensively tested and trialled before being used to make sure they are safe (not toxic), effective, and to work out the dose required.

Drugs are tested using computer models, in the laboratory on cells and tissues, and on live animals before being tested on human volunteers. Testing on animals raises ethical issues, but new technologies are being developed instead.

Clinical trials involve healthy volunteers and patients. Very low doses of the drug are given at the start of a trial. If the drug is found to be safe, further trials are carried out to determine the optimum dose for the drug.

In some **double blind** trials, some patients are given a **placebo**, which does not contain the drug. Neither the doctors nor the patients know who has received a placebo and who has received the drug until the trial is complete. This prevents **bias**.

It might not be considered ethical to give patients placebos as they still need treatment, so placebos are not often used.

An **open-label** drugs trial is when the doctor and the patients know the name of the drug they are taking.

Sometimes tests fail. This happened with a drug called **thalidomide** that was developed as a sleeping pill. It was found to be effective at treating morning sickness; however, it had not been tested for this use. This drug caused severe limb abnormalities in babies born to mothers taking the drug. It was then banned.

It is now being used to treat leprosy and some forms of cancer. As a result of the issues with thalidomide, drug testing has become more rigorous.

SUMMARY

- Vaccines help to prevent the spread of disease.
- Antibiotics are used to treat bacterial infections.
- Drugs must be tested to make sure they are safe, effective, and to work out the correct doses needed.

QUESTIONS

QUICK TEST

1. How do bacteria and viruses become resistant to antibiotics?

2. Which three diseases does the MMR vaccine treat?

3. Name the fungi that is used to produce the antibiotic penicillin.

EXAM PRACTICE

1. Laura has influenza. The doctor won't prescribe her with antibiotics. Explain why. **(2 marks)**

2. Bethany has had the flu vaccination.

   a) What is a vaccine? **(2 marks)**

   b) Explain why this vaccine doesn't prevent her from getting a cold. **(2 marks)**

   c) List two benefits of being vaccinated. **(2 marks)**

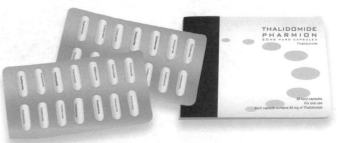

Drugs

Drugs are chemicals that alter the way the body works, causing changes in behaviour and possible addiction. Drugs can be beneficial or harmful.

Drugs can become **addictive** – people become dependent on them and suffer from withdrawal symptoms when they stop taking them.

Some drugs taken for **recreational** use are **illegal**, but other drugs, such as alcohol and cigarettes, are **legal**. The overall impact on health from legal drugs can be greater as more people take them than the illegal drugs.

Drugs fall into the following main groups:

1. **Depressants**

   Depressants slow down the activity of the brain by blocking the transmission of nerve impulses. This seriously alters reaction times and affects a person's judgement of speed and distance when driving. Examples include alcohol, solvents, tranquillisers and sleeping pills, e.g. temazepam.

2. **Painkillers**

   Painkillers suppress pain by blocking nerve impulses. Paracetamol, aspirin, heroin and morphine are examples of painkillers. Heroin and cocaine are very addictive.

3. **Hallucinogens**

   Hallucinogens can cause hallucinations (seeing or hearing things that do not exist) by distorting how we perceive our senses. Examples include Ecstasy, LSD and cannabis. **Cannabis** is an illegal drug and its smoke contains chemicals that may cause mental illness in some people. **Ecstasy** (MDMA) results in a smaller volume of less dilute urine because it increases ADH production. Ecstasy also blocks synapses in the brain where serotonin is removed, so serotonin levels increase.

4. **Stimulants**

   Stimulants increase the speed of reactions and neurotransmission at the synapse, making you feel more alert and awake. Examples include amphetamines, cocaine, Ecstasy and the less harmful **caffeine** in tea and coffee.

5. **Performance-enhancing drugs**

   Some athletes use drugs to enhance performance. Some of these drugs are banned by law and some are legally available on prescription, but all are prohibited by sporting regulations. Examples include:

   - stimulants that boost bodily functions such as heart rate
   - **anabolic steroids** that stimulate muscle growth.

Alcohol

Alcohol is a depressant, absorbed through the gut and taken to the brain in the blood. In the short term it can cause blurred vision, lowering of inhibitions and slowing of reactions. There is a legal limit for the level of alcohol in the blood for drivers and pilots for this reason.

In the long term, alcohol can damage neurones in the brain, causing irreversible brain damage. The liver breaks down alcohol. In excess, alcohol has a very damaging effect on the liver, causing a disease called **cirrhosis**.

Alcohol consumption results in a greater volume of more dilute urine produced because it suppresses ADH. This can lead to dehydration that can adversely affect health.

Smoking

Tobacco contains many harmful chemicals. **Nicotine** is an addictive substance and a mild stimulant. **Tar** is known to contain carcinogens that contribute to cancer. **Carbon monoxide** prevents the red blood cells from carrying oxygen by combining with haemoglobin, which increases heart rate and blood pressure. Particulates accumulate in the lung tissue.

If pregnant women smoke, then carbon monoxide deprives the foetus of oxygen and can lead to a lower birth mass. Some diseases caused by smoking include emphysema, bronchitis, heart and blood vessel problems and cancer (mouth, throat, oesophagus and lung).

SUMMARY

- Drugs can be depressants, painkillers, hallucinogens, stimulants or performance enhancing.
- Alcohol is a depressant and can damage the brain and liver.
- Tobacco smoke contains tar, nicotine, carbon monoxide and particulates.

Drugs and the Law

Drugs are classified in law as class A, B or C. Class A drugs, such as heroin, carry the most severe penalties if a person is caught in possession of them. Supply of the drug can lead to life imprisonment.

Class C drugs are considered the least dangerous with the lightest penalties given.

Cannabis is an illegal drug. Many people argue whether it should be a class B or C drug, or even if it should be made legal.

The cannabis debate still continues about whether it is harmful, addictive, or leads onto the use of harder drugs such as heroin. At present, health professionals cannot agree.

QUESTIONS

QUICK TEST

1. How do sedatives slow down the activities in the brain?

2. What effect could cannabis cause in some people?

3. What do anabolic steroids do?

EXAM PRACTICE

1. A group of students wanted to investigate the effect of caffeine in cola on their reaction time.

   Write a plan that would allow the students to carry out their investigation.

   The quality of your written communication will be assessed in your answer to this question.

   (6 marks)

Inheritance and Genetics

Inside nearly all cells is a nucleus that contains instructions that control all our inherited characteristics.

The information that results in plants and animals having similar characteristics to their parents is carried by genes, which are passed on in sex cells (gametes) from which the offspring develop.

The **nucleus** of a cell contains chromosomes. **Chromosomes** carry genes that control the characteristics of the body. There are 46 chromosomes in human body cells which are in pairs (one set from each parent). **Gametes** only contain 23 chromosomes.

Each chromosome is made up of a long strand of DNA. A **gene** is a section of DNA. Different genes control the development of different characteristics in an organism. DNA contains coded information for the production of proteins. These proteins determine how cells function.

Genetics

Genetics is the study of how information is passed on through generations.

Gregor **Mendel** discovered the principles behind genetics by studying the inheritance of a single factor (colour) in pea plants. It wasn't until the development of powerful microscopes after Mendel's death that the basic facts of cell division, sexual reproduction and consequently inheritance were understood.

Definitions

- **Gene**: the unit of inheritance carried on chromosomes that controls characteristics. Many characteristics are determined by several genes working together.
- **Alleles**: alternative forms of a gene.
- **Genotype**: the alleles the organism carries.
- **Phenotype**: what the organism physically looks like because of the alleles expressed.
- **Recessive**: only has an effect in the homozygous recessive genotype.
- **Dominant**: means it is the stronger allele and has an effect in the heterozygous genotype by suppressing the recessive allele.

- If an organism has both alleles the same, they are **homozygous dominant** (e.g. BB) or **homozygous recessive** (e.g. bb).
- If an organism has different alleles, they are **heterozygous** (e.g. Bb).

A Worked Example

The allele for brown eyes is **dominant** (**B**) to the **recessive** blue eyes allele (**b**).

Letters are used to represent alleles: upper case for dominant characteristics and lower case for recessive characteristics.

If the mother and father are **heterozygous** for eye colour, then they have the genotype **Bb**. What colour eyes will their children have?

We can show the possible outcomes using a **genetic diagram**.

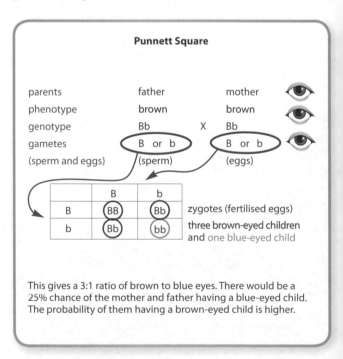

Punnett Square

parents	father	mother
phenotype	brown	brown
genotype	Bb	X	Bb
gametes (sperm and eggs)	B or b (sperm)	B or b (eggs)

	B	b
B	BB	Bb
b	Bb	bb

zygotes (fertilised eggs)
three brown-eyed children and one blue-eyed child

This gives a 3:1 ratio of brown to blue eyes. There would be a 25% chance of the mother and father having a blue-eyed child. The probability of them having a brown-eyed child is higher.

Family Pedigree Charts

Pedigree charts, or family trees, can be used to show the way characteristics of individuals related to each other pass from one generation to the next.

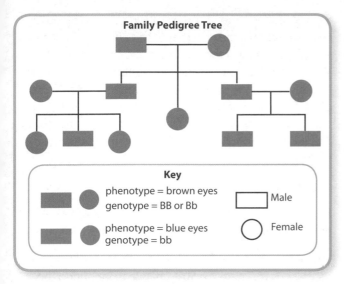

Family Pedigree Tree

Key

phenotype = brown eyes
genotype = BB or Bb

phenotype = blue eyes
genotype = bb

☐ Male

◯ Female

Inheritance

To inherit characteristics from the parent DNA, reproduction needs to take place. There are two types of reproduction.

Asexual reproduction does not involve the fusion of gametes (sperm and egg) and involves only one parent. As there is no mixing of genetic information there is no variation in the offspring. The offspring have exact copies of the parental genes – i.e. they are **clones**. Any difference between the clones is likely to be due to environmental factors. Clones of animals occur naturally when cells of an embryo separate to form **identical twins**.

Sexual reproduction involves fertilisation and two parents. The male and female gametes' nuclei fuse and the genes are passed on. The offspring are not genetically identical and the mixture of genetic information from two parents causes variation in the offspring. Each gamete is **haploid**, i.e. it contains half the number of chromosomes (23). Fertilisation restores the number of chromosomes to 46 producing a **diploid** zygote.

Sexual reproduction gives rise to **variation** because when the gametes fuse, one of each pair of alleles comes from each parent. However, brothers and sisters from the same parent may look different because they inherit a different combination of alleles.

SUMMARY

● The nucleus contains chromosomes made of DNA that contains genes.

● Different versions of genes are called alleles and can be dominant or recessive.

● Punnett squares and family pedigree diagrams help to predict the probability of inheritance.

● Reproduction can be sexual or asexual.

QUESTIONS

QUICK TEST

1. What does homozygous mean?

2. What is asexual reproduction?

3. What type of cell is haploid?

EXAM PRACTICE

1. Tongue rolling is dominant to non-tongue rolling. Chris has the genotype Rr and Helen has the genotype Rr.

   a) What are the possible genotypes for tongue rolling and non tongue rolling? **(1 mark)**

   b) What are the possible genotypes of their children? Use a genetic diagram in your answer. **(2 marks)**

   c) What is the probability of Chris and Helen having a non-tongue rolling child? **(1 mark)**

2. Megan has two black cats. They breed and produce kittens. Explain why some of the kittens have grey fur and some have black fur.

   The quality of your written communication will be assessed in your answer to this question.

   (6 marks)

Genetic Disorders

Genes pass on characteristics from one generation to the next. Sometimes 'faulty' genes are inherited that cause genetic diseases.

Cystic Fibrosis

Cystic fibrosis is a common inherited disorder of the cell membranes. It is caused by a **recessive allele** (**f**). People who have the genotype (Ff) are said to be **carriers**, with no ill effects. Only people who have the genotype (ff) are affected. The recessive gene is inherited from both parents.

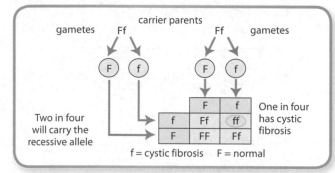

Cystic fibrosis sufferers produce large amounts of thick, sticky **mucus** that can block air passages and digestive tubes. This causes difficulty in breathing and absorbing food. The mucus also encourages bacteria to grow, which can cause chest infections.

There is still no cure for cystic fibrosis. Treatment involves **physiotherapy** to remove some of the mucus, and strong **antibiotics** to treat infections.

Sickle Cell Anaemia

Sickle cell anaemia is caused by a **recessive allele** (**a**). Sufferers have the genotype (aa).

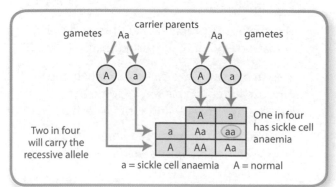

Sickle cell affects the **blood** by causing the red blood cells to have an abnormal shape. This affects the oxygen-carrying capacity of the blood and the sickle cells get stuck in capillaries. It is painful and sufferers usually die at an early age. There is no cure, and even though sufferers die before they can reproduce, the disease has not disappeared, especially in malarial regions such as Africa. This is because **carriers** with the genotype (**Ss**) are more likely to survive malaria.

Huntington's Chorea

Huntington's chorea is caused by a **dominant allele** (**H**). All **heterozygous people are sufferers** (**Hh**).

The only people free from the disease are homozygous recessive (hh). There is a 50% chance of inheriting the disease if just one parent is a carrier.

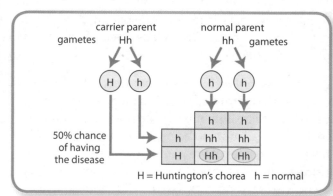

Symptoms of Huntington's chorea include tremors, clumsiness, memory loss, inability to concentrate and mood changes. There is no cure and onset of the disease is late. Consequently, many sufferers have already had children and passed on the dominant allele before they realise they have it.

Polydactyly

People with **polydactyly** have extra fingers or toes. Polydactyly is caused by a **dominant allele**, so it can be passed on even if only one parent has the disorder.

Protein Synthesis

The development, structure and function of an organism are determined by which proteins are made in the body in a process called **protein synthesis**. DNA carries the codes for proteins on genes. The order of **bases** in a section of DNA decides the order of amino acids in the protein.

Stage 1 – Transcription

1. The DNA molecule unwinds to expose a sequence of bases (a gene).
2. The sequence of bases is copied and forms a complementary molecule called messenger RNA (**mRNA**) in the nucleus. Genes themselves do not leave the nucleus.
3. The mRNA moves out of the nucleus and carries the copy of the gene onto a structure called a **ribosome**.

Stage 2 – Translation

4. A second type of RNA called transfer RNA (**tRNA**) attaches to the mRNA, bringing with it a particular **amino acid**.
5. The sequence of amino acids is determined by the order of bases on the mRNA.
6. The amino acids are linked in a certain order and form the polypeptide (protein) required.

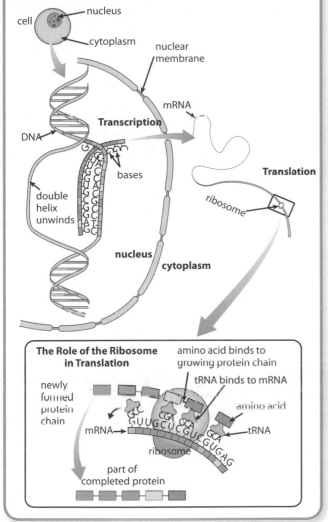

A sequence of three bases (codon) on the tRNA will only match against a particular sequence on the mRNA.

Although all body cells contain the same genes, many genes in a cell are not active (switched off) because the cell only produces the specific proteins it needs so only needs those genes switched on.

SUMMARY

- Sometimes 'faulty' genes are inherited that can lead to genetic disorders.
- Cystic fibrosis and sickle cell anaemia are both caused by recessive alleles.
- Genetic diagrams can show the probability of offspring being sufferers.

QUESTIONS

QUICK TEST

1. Describe the genotype of a sufferer of cystic fibrosis.
2. What is the benefit of being a sickle cell carrier?
3. What type of allele causes polydactyly?

EXAM PRACTICE

1. The diagram shows a family tree for the inheritance of cystic fibrosis.

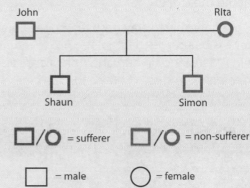

a) What evidence is there that cystic fibrosis is coded for by a recessive allele? **(2 marks)**

b) Are cystic fibrosis carriers heterozygous or homozygous? **(1 mark)**

c) List two symptoms of cystic fibrosis. **(2 marks)**

Genetic Engineering

Genetic engineering is the process in which genes from one organism are removed using enzymes and inserted into the cells of another.

Genes can also be transferred to the cells of animals, plants or microorganisms at an early stage in their development so that they develop with desired characteristics.

Crops that have had their genes modified by having new genes inserted are called **genetically modified** crops (GM crops). Examples of GM crops include ones that are resistant to insect attack or to herbicides. GM crops generally show increased yields.

Benefits of Genetic Engineering

Industry, medicine and agriculture all benefit from genetic engineering. For example, plants that are resistant to pests and diseases can grow in less desirable conditions. Wheat and other crops have been developed to contain nitrogen-fixing bacteria like legumes so they are able to produce proteins without the need for fertilisers. Pesticide-resistant plants, such as soya beans, have been developed so that pesticides can be applied to whole fields without damaging the crop.

Fruits are now able to stay fresh for longer. Beta carotene is being added to golden rice to reduce vitamin A deficiency in humans. Animals are also engineered to produce chemicals in their milk such as drugs and human antibodies.

Risks of Genetic Engineering

There is concern that GM crops damage human health by causing allergies. There is also concern for the nutritional quality of GM crops. It is argued that GM crops damage the environment, and there is the risk of crops transferring their herbicide resistance to weeds.

Genetic engineering is seen by some as manipulating the 'stuff of life'. Is it morally and ethically wrong to interfere with nature? But, on the other hand, could it help food shortages in some countries and save lives?

Gene Therapy

It may be possible to treat inherited diseases such as **cystic fibrosis**. Sufferers could be cured if the correct gene could be inserted into their body cells, but the cells that need the gene are in many parts of the body.

Gene therapy is a technique for correcting defective genes responsible for disease development. In most studies a viral vector carrying a 'normal' gene is inserted into the patient's cells to replace an 'abnormal' disease-causing gene.

Genetic Screening

Adults, children and embryos can be screened for the alleles that cause genetic disorders. Embryos can be tested during IVF. Individuals can be tested before prescribing drugs.

Risks associated with testing adults and foetuses for alleles that cause genetic disorders include a risk of miscarriage and inaccurate results (false positives and false negatives). Difficult decisions may also need to be made, such as whether to have children at all, whether to terminate a pregnancy, or whether other family members should be informed.

DNA technology can be used in genetic testing:
1. A DNA sample is isolated, usually from white blood cells.
2. A gene probe is produced, labelled with a fluorescent chemical.
3. The gene probe is added to the DNA sample.
4. UV is used to detect the marker and therefore indicates the position of the gene or the presence of a specific allele in the DNA sample.

Human Genome Project

The Human Genome Project was finalised in 2003. It took scientists from all around the world 10 years to complete. It aimed to identify all the genes in human DNA and study them. The benefits included improved diagnosis for diseases and earlier detection of genetic diseases such as breast cancer and Alzheimer's.

The use of DNA in **forensic science** was also improved. It is possible to identify suspects, clear the wrongly accused, identify paternity and match organ donors with recipients. A person's '**DNA fingerprint**' or 'DNA profile' is unique (apart from identical twins). A sample is taken from the body tissue or fluid (hair, blood or saliva). Enzymes cut the DNA into fragments and they are separated using a process called **electrophoresis**.

DNA profiling can help to identify the presence of certain genes that may be associated with a particular disease. There are concerns over who should have access to this information. For example, insurance companies could use this information to see how long you are expected to live, and may sell you products at a higher rate or refuse to sell them to you at all.

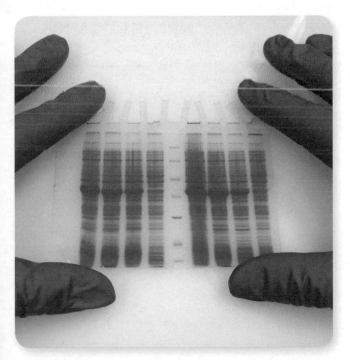

SUMMARY

- **Genetic engineering occurs when genes from one organism are placed into the DNA of another.**

- **GM crops have been developed for resistance to pests or herbicides.**

- **Gene therapy aims to correct defective genes that cause diseases.**

QUESTIONS

QUICK TEST

1. What is genetic engineering?

2. What is gene therapy?

3. What does 'genetically modified' mean?

EXAM PRACTICE

1. DNA can be isolated and separated for use in forensic science.

   a) What is the name of the process used to separate out DNA into fragments? **(1 mark)**

   b) Outline what the Human Genome Project was. **(1 mark)**

   c) Give three benefits that have risen from the completion of the Human Genome Project. **(3 marks)**

2. GM crops can be considered very controversial. Give two arguments for the use of GM crops, and two against GM crops. **(4 marks)**

3. Genetic screening can be carried out on a foetus.

   a) Give an advantage and a disadvantage of genetic screening for parents. **(2 marks)**

   b) Give an ethical consideration. **(1 mark)**

Manipulating Life

Selective breeding is where humans try to improve animals and plants by breeding the best individuals together and hoping that it is successful. It can also be called **artificial selection**.

Selective Breeding in Animals

Dogs have been selectively bred over many years to produce the variety of breeds that we have today. Cows have been selectively bred to produce a greater quantity of milk or better quality beef.

New techniques have been developed to produce more offspring in a shorter space of time, e.g. embryo transplantation.

Embryo transplantation is where cells from a developing animal embryo are split apart before they become specialised. The identical embryos are then transplanted into surrogate mothers.

Adult cell cloning, another technique, is more complex and is shown in the diagram below.

Under the carefully controlled conditions of cloning, it is possible to reactivate (switch on) inactive genes in the nucleus of a body cell to form cells of all tissue types.

Dolly the sheep was the first mammal cloned in 1996. She died prematurely in 2003 following a lung infection, but her early death fuelled the debate about the long-term health problems of clones.

Scientists are looking at ways of using genetically engineered animals to grow replacement organs for humans. This poses many ethical concerns, not least the problems of rejection.

Selective Breeding in Plants

Selectively bred individuals may not always produce the desired characteristics if it involves sexual reproduction. With plants this can be overcome by producing clones, using **asexual reproduction**.

Gardeners can produce identical plants, quickly and cheaply, by taking **cuttings** from an original plant. The cuttings are dipped in hormone powder and then grown into new plants. It is possible to mass produce plants that may be difficult to grow from seed.

Tissue culture is when small groups of cells from a part of a plant are grown into a new plant using a special growth medium containing hormones. All the plants are identical to the original so you can be sure of the characteristics. It is possible to grow large numbers from just a small piece of tissue. **Aseptic techniques** must be followed.

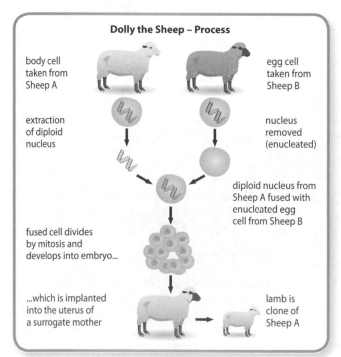

Dolly the Sheep – Process

body cell taken from Sheep A

extraction of diploid nucleus

egg cell taken from Sheep B

nucleus removed (enucleated)

diploid nucleus from Sheep A fused with enucleated egg cell from Sheep B

fused cell divides by mitosis and develops into embryo...

...which is implanted into the uterus of a surrogate mother

lamb is clone of Sheep A

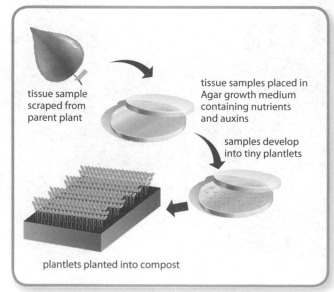

tissue sample scraped from parent plant

tissue samples placed in Agar growth medium containing nutrients and auxins

samples develop into tiny plantlets

plantlets planted into compost

Plant cloning by tissue culture is easier than cloning animals because many plant cells retain the ability to differentiate, unlike animal cells which usually lose this ability at an early stage.

Many plants reproduce asexually on their own, such as strawberry plants that produce runners. Other examples are potatoes, onions and *Chlorophytum* (spider plant). Plants also reproduce sexually, attracting insects for pollination.

Spider Plant Reproducing Asexually

Problems with Selective Breeding

If plants or animals are continually bred from the same plants and animals, then the offspring will all be very similar. This can lead to inbreeding and cause health problems. A change in the environment, or a new disease, may mean that the new plants or animals will not be able to cope with the change and die out. There is a lack of genetic variation. The gene pool contains less alleles, reducing further selective breeding options.

SUMMARY

- Humans use selective breeding to produce plants and animals with desired characteristics.
- Animals can be reproduced using embryo transplant and adult cell cloning techniques.
- Plants can be reproduced using cuttings and tissue culture techniques.

QUESTIONS

QUICK TEST

1. What type of reproduction produces clones?

2. Are calves that are produced by embryo transplant clones?

3. Why is Dolly the sheep famous?

EXAM PRACTICE

1. a) From what type of cell was the DNA extracted to clone Dolly? **(1 mark)**

   b) Looking at the 'Dolly the Sheep' diagram on page 44, explain why the nucleus was removed from Sheep B's cell. **(1 mark)**

   c) What is the name for this type of cloning method? **(1 mark)**

2. Donna is a gardener and she wants to make copies of her show-winning plants.

   a) List two techniques she could use to produce clones of her plant. **(2 marks)**

   b) Donna has a very small budget. Which technique would you suggest she uses? Explain your answer. **(2 marks)**

The Heart

The circulatory system consists of the heart, blood vessels and blood. It transports substances around the body, some of which are needed for respiration.

There are four chambers of the heart (left and right **atria** and **ventricles**). The heart is a double pump. The left side, with thicker walls, pumps **oxygenated** blood at high pressure out of the aorta to other arteries delivering substances around the body. The right side of the heart carries **deoxygenated** blood to the lungs to be oxygenated. The heart contains valves to prevent the backflow of blood.

The heart is made of cardiac muscle, with its own blood supply of **coronary arteries** that supply the heart with oxygen and nutrients. These arteries can become blocked. Treatment of blocked arteries involves bypass surgery. **Stents** can be used to keep narrow arteries open. Biomedical engineering involves repairing or replacing damaged heart valves and developing artificial pacemakers that can be fitted if the heart beats irregularly.

Heart transplants can be problematic because there is a shortage of donor hearts, and donor hearts have to be a correct match in size, age and tissue type or **rejection** can occur. Even if the donor heart is a match, there is still a risk of rejection. The patient receiving the heart has to take anti-rejection drugs for the rest of his / her life.

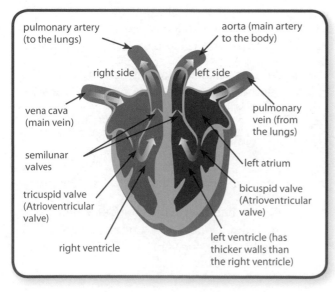

Blood Vessels

Veins carry **deoxygenated** blood **back to the heart** from the body at **low pressure**. They have thin walls, a large lumen and **valves** to prevent the blood flowing backwards.

Arteries carry oxygenated blood **away from the heart** towards the body at **high pressure**. They have very thick, muscular elastic walls to withstand the high pressure.

Capillaries are only one cell thick and have very thin, permeable walls. This allows oxygen and nutrients to diffuse out of them into the cells, and waste substances from the cells to diffuse into the blood. Capillaries form an extensive network so that every cell is near to a capillary that is carrying blood.

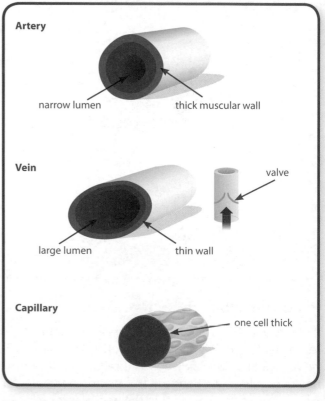

The Cardiac Cycle

The heartbeat follows a series of events called the **cardiac cycle**. It consists of the atria and ventricles contracting (systole) and then relaxing (diastole).

1. The heart is relaxed and both atria fill up with blood from the veins. At this point the valves between the atria and ventricles are open.

2. The atria contract and force the valves open so that blood flows into the ventricles.

3. The valves close and a fraction of a second later the ventricles contract, forcing the blood out of the aorta and pulmonary artery.

4. The whole heart relaxes and fills up with blood again.

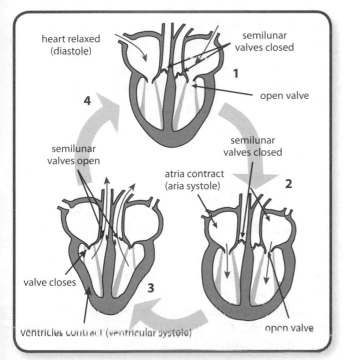

Control of the Cardiac Cycle

Each heartbeat is controlled by a **pacemaker**, a special area of tissue in the atrium wall called the SAN. The SAN produces a small electric current that passes to another area of cells in between the two atria called the AVN. The AVN passes this current along bundles of fibres between the ventricles. Fibres then spread the current upwards through the ventricle walls.

The heart beats faster when a hormone called **adrenaline** is secreted, or during exercise where a rise in blood carbon dioxide is detected by the brain which stimulates the SAN via nerve impulses.

Tissue Fluid

The fluid that surrounds cells is called **tissue fluid**. It assists in the exchange of substances by diffusion between capillaries and tissues. Substances that diffuse include oxygen, carbon dioxide, glucose and urea.

SUMMARY

- **The circulatory system consists of the heart, blood vessels and blood.**
- **The heart is a double pump.**
- **Arteries, veins and capillaries are the blood vessels.**
- **A pacemaker controls the cardiac cycle.**

QUESTIONS

QUICK TEST

1. Which blood vessel carries blood way from the heart?

2. Why do arteries have elastic, muscular walls?

3. What do we call the pacemaker of the heart?

EXAM PRACTICE

1. Describe the sequence of events in the cardiac cycle.

   You will be assessed on the quality of your written communication in your answer to this question.
   (6 marks)

2. Parts of the heart or the whole heart can be replaced.

   a) What problems are there with finding donor hearts? **(2 marks)**

   b) Why are transplants not always successful? **(1 mark)**

The Circulatory System

We need a transport system to deliver food and oxygen to our cells and to remove carbon dioxide and waste substances quickly and efficiently.

A **double circulatory system** means that the blood passes through the heart twice on one journey around the body.

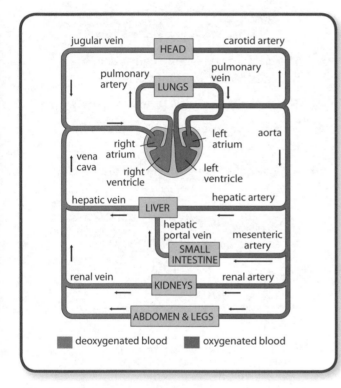

jugular vein • carotid artery • HEAD • pulmonary artery • LUNGS • pulmonary vein • right atrium • left atrium • aorta • vena cava • right ventricle • left ventricle • hepatic vein • hepatic artery • LIVER • hepatic portal vein • mesenteric artery • SMALL INTESTINE • renal vein • renal artery • KIDNEYS • ABDOMEN & LEGS

■ deoxygenated blood   ■ oxygenated blood

All vertebrates have a **closed** system meaning that blood never leaves the vessels.

Some animals, such as insects, have an **open** system whereby blood fluid bathes their organs. Muscle movement causes the fluid to move around the body.

Some small organisms do not have a circulatory system at all, but rely on diffusion to supply them with nutrients.

Galen and Harvey

In the 2nd Century, a Greek physician called **Galen** knew that blood vessels carried blood and that there was a difference between venous blood that was dark and arterial blood that was bright red. He knew they had separate functions but did not fully understand the role of the heart in circulation.

Finally a scientist, William **Harvey**, carried out experiments and in 1628 discovered the workings of the human circulatory system.

Under Pressure

Heart rate can be measured by recording the pulse rate, taken at places such as your wrist, neck or behind the knee. Blood is put under pressure in the arteries so that it can reach all parts of the body. The heart muscles contract to force the blood out of the arteries.

The total amount of blood pumped by the heart per minute is the **cardiac output**. Cardiac output can be calculated using the following equation:

$$\text{Cardiac output} = \text{stroke volume} \times \text{heart rate}$$

Stroke volume is the volume of blood pumped in one beat.

Blood pressure is a measure of the pressure of the blood on the wall of the artery and consists of a **diastolic** measurement (when the heart is at rest) and a **systolic** measurement (when the heart contracts).

Blood pressure is given as two numbers. The higher value is the systolic measurement and is the maximum pressure exerted by the heart muscle pushing the blood through the blood vessels. The lower value is when the heart is relaxed and is the resistance to blood flow in the blood vessels.

Blood pressure is measured in mm Hg (millimetres of mercury) using an instrument called a sphygmomanometer. Abnormal blood pressure can indicate problems with the circulatory system.

'Normal' measurements for heart rate and blood pressure are given within a range because individuals vary due to a range of factors, e.g. age, sex, diet, exercise, weight, smoking, alcohol intake and stress.

Low blood pressure can cause poor circulation and fainting. High blood pressure can be harmful as it puts a strain on your heart and could cause an artery to burst. If this happens in the brain it is called a stroke.

Heart Conditions and Diseases

There are many heart conditions and diseases including irregular heartbeats, holes in the heart, damaged or weak valves, coronary heart disease and heart attacks.

A **hole in the heart** can result in less oxygen in the blood because the blood would be able to move directly from one side of the heart to the other through the hole. Surgery can correct this condition. Unborn babies can have a hole in the heart because they do not need a double circulatory system; after birth the hole closes up.

Fatty deposits can build up in the blood vessels supplying the heart muscle. This can cause a **heart attack**. Heart disease is usually caused by lifestyle factors and/or genetic factors.

Lifestyle factors that can increase the risk of heart disease include a poor diet (eating high levels of salt and saturated fat), stress, high blood pressure, smoking and drug abuse. Regular, moderate exercise helps to reduce the risk of heart disease.

ECG

An electrocardiogram (**ECG**) can be used to investigate heart action. The electrical activity of the heart is read by electrodes placed at specific points of the body. A normal trace is shown below.

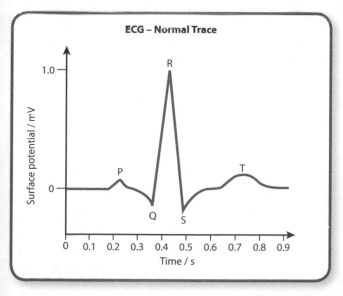

ECG – Normal Trace

SUMMARY

- Circulatory systems can be open or closed, and single or double.
- Humans have a double circulatory system.
- Blood pressure is measured using a sphygmomanometer.
- An ECG can detect problems with the circulatory system.

QUESTIONS

QUICK TEST

1. Do humans have a closed or open circulatory system?

2. Who is credited with discovering the workings of the human circulatory system?

3. List three factors that can increase a person's chance of heart disease.

EXAM PRACTICE

1. Fish only have a single circulatory system.

   a) What are the advantages of having a double circulatory system? **(2 marks)**

   b) Why do we need a transport system? **(2 marks)**

2. A person has a blood pressure of 120 over 80.

   a) What does the 120 mean? **(1 mark)**

   b) What advice would you give someone with high blood pressure in order to help them lower it? **(2 marks)**

The Blood

Blood is a tissue that transports food and oxygen to cells and removes waste products. Blood is carried in vessels called veins, arteries and capillaries.

Blood consists of **red blood cells**, **white blood cells** and **platelets** suspended in a fluid called **plasma**.

Red blood cells carry oxygen to all the cells of the body.

Red blood cells have adapted so that they can do this efficiently:

- They contain a substance called **haemoglobin**. In the lungs it combines with oxygen to form **oxyhaemoglobin**.
- They have **no nucleus** so there is more room for haemoglobin.
- They are **small** and **flexible** to pass through narrow blood vessels.
- Their shape is a small **biconcave** disc, which gives a maximum surface area to volume ratio for absorbing oxygen.

White blood cells help defend the body against disease. They have a large nucleus and their shape varies.

Platelets are fragments of cells with no nucleus, whose function is to help with the coagulation (blood clotting) process. Haemophilia is a condition whereby blood does not easily clot due to the absence of certain blood clotting factors.

Plasma consists of mainly water, with dissolved substances such as soluble food, salts, carbon dioxide, urea, hormones, antibodies and plasma proteins.

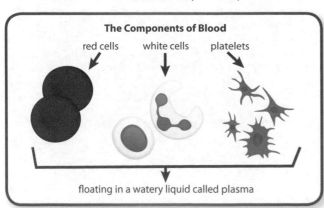

The Components of Blood

red cells   white cells   platelets

floating in a watery liquid called plasma

Blood Groups

There are four different blood groups: A, B, AB and O. These groups are further subdivided into Rhesus positive and negative.

A blood transfusion is the introduction of blood from one person (the **donor**) into the circulation of another (the **recipient**). It is important with a transfusion that the donor and recipient's blood is matched, or coagulation will occur.

Antibodies in plasma will attack antigens on the surface of red blood cells if they are incompatible. This means that…

- blood group A can only receive blood from type A or O
- blood group B can only receive blood from type B or O
- blood group AB can receive blood from type A, B, AB or O, but can only donate to other AB recipients
- blood group O can only receive blood from type O, but can donate to all groups.

Blood group O is known as the universal donor and blood group AB as the universal recipient.

Blood group	Antigens on red blood cell	Agglutinin (antibodies) in blood plasma	
A	Type A	b	Antibody b will bind to antigen B
B	Type B	a	Antibody a will bind to antigen A
AB	Type AB	neither	Will not bind to antigens as no antibodies in plasma
O	Neither	a and b	Antibodies a and b will bind to A and B antigen

Blood Clotting

Blood forms clots normally at cuts and sometimes inside blood vessels. Blood clotting does not just involve platelets, but proteins and enzymes. A series of events ends with a protein called fibrin and platelets forming a platelet plug over the cut.

Anticoagulant drugs such as warfarin, heparin and aspirin can help prevent the blood from clotting so easily. Vitamin K, found in the body and in leafy green vegetables, helps the blood clotting process.

SUMMARY

- Blood is made up of four parts, each carrying out different functions.
- Red blood cells transport oxygen using haemoglobin.
- White blood cells are part of our immune system, and help defend the body against disease.
- There are four different blood groups.
- During transfusions it is important that blood groups match.

QUESTIONS

QUICK TEST

1. Which blood group can donate blood to any other blood group?

2. Which blood group can receive blood from any other blood group?

3. What does coagulation mean?

EXAM PRACTICE

1. Complete the table by…

   a) listing three features of red blood cells that make them good at carrying oxygen. (3 marks)

   b) explaining how these features benefit the red blood cell in its role. (3 marks)

Feature	Benefit in the Role of Carrying Oxygen

Energy, Biomass and Photosynthesis

The ultimate source of energy in the environment is the energy from sunlight, which radiates and is harnessed by green plants through **photosynthesis**.

Pyramid of Numbers

A **food chain** or **food web** shows us the direction of energy transfer and therefore 'who eats who'. A **pyramid of numbers**, however, shows us how many organisms are involved at each stage in a food chain. At each level of a food chain (**trophic level**) the number of organisms generally gets less.

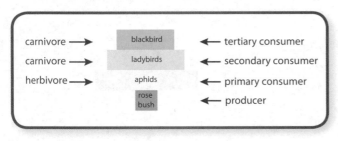

Pyramid of Biomass

The mass of living material (**biomass**) at each trophic level in a food chain is less than it was at the previous stage and is shown in a **pyramid of biomass**. A pyramid of biomass must be drawn to scale.

Pyramids can be difficult to construct because organisms may belong to more than one trophic level and there could be problems with measuring dry biomass.

Loss of Energy

Food chains rarely have more than four or five links. The final organism is only getting some of the energy that was produced at the beginning of the food chain.

Green plants and other photosynthetic organisms, such as algae, absorb their energy from the Sun. Only a small fraction of this energy is converted into glucose during photosynthesis. Some energy is lost to decomposers as plants shed their leaves, seeds or fruit. The plant uses some energy during respiration and growth. The plant's biomass increases, which provides food for herbivores.

At each stage in the food chain energy is used in repair, maintenance and growth of new cells, and energy is lost in waste and heat to keep warm. Much of this energy is eventually transferred to the surroundings.

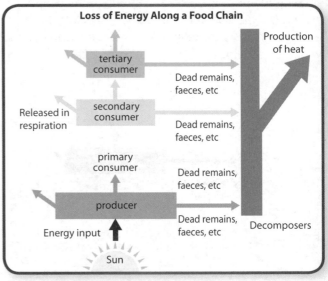

Photosynthesis

The **leaf** is the organ of **photosynthesis**. Photosynthesis is a chemical process (controlled by enzymes) that plants and algae use to make glucose. Oxygen is released as a by-product.

Within the plants or algae, glucose may be…

- ⚫ respired to provide energy
- ⚫ converted to insoluble starch for storage
- ⚫ used to make fats and oils (for storage), cellulose (for structure) and proteins that make up the body of plants.

The equation for photosynthesis is:

$$\text{carbon dioxide} + \text{water} \xrightarrow[\text{chlorophyll}]{\text{light}} \text{glucose} + \text{oxygen}$$

$$6CO_2 + 6H_2O \rightarrow C_6H_{12}O_6 + 6O_2$$

Photosynthesis is a two-stage process:

1. Light energy is used to split water, releasing oxygen gas and hydrogen ions.
2. Carbon dioxide gas combines with the hydrogen ions to make glucose.

The leaf has many features that enable it to carry out photosynthesis efficiently. Leaves:

- are **flat** with a **large surface area** to absorb sunlight
- are **thin** so carbon dioxide can reach the inner cells easily
- have **plenty of stomata** in the lower surface for gas exchange by diffusion
- have **plenty of veins** to support the leaf and carry substances to and from all the cells in the leaf and plant
- contain **chlorophyll** in chloroplasts to absorb light.

Factors Affecting the Rate of Photosynthesis

We can measure the rate of photosynthesis by how much oxygen is produced in a given time. The limiting factors that affect the rate of photosynthesis are:

- light intensity
- concentration of carbon dioxide
- temperature.

At any given time, any one of these factors could be limiting the rate of photosynthesis. Usually the rate of photosynthesis is limited by the temperature being too low, as is the case for plants not normally grown in Britain.

Greenhouses help maintain a high enough temperature for optimum growth conditions – the amount of carbon dioxide and light can be maximised too.

SUMMARY

- **Energy is lost along a food chain, which limits the number of trophic levels.**
- **Plants and algae carry out photosynthesis to produce glucose and release oxygen.**
- **The leaf is adapted to carry out photosynthesis.**
- **Limiting factors affect the rate of photosynthesis.**

QUESTIONS

QUICK TEST

1. What type of pyramid takes into account the number of organisms in a food chain?

2. What is a primary consumer?

3. Why are leaves flat?

EXAM PRACTICE

1. Edward investigated the levels of carbon dioxide in different samples of pond water. He used three samples:

   A – Open test tube with indicator and pond water only

   B – Sealed test tube with indicator, pond water and pond weed

   C – Sealed test tube with indicator, pond water and pond weed, covered in tinfoil

   Edward used the indicator sodium hydrogencarbonate. In high levels of CO_2, sodium hydrogencarbonate turns yellow, in atmospheric levels it is red, and in low levels of CO_2 it turns purple.

   a) What would you expect the colour of each indicator to turn in test tubes A–C? Explain your answer.   **(3 marks)**

   b) How can Edward make his results more reliable?   **(1 mark)**

   c) What is the independent variable in his investigation?   **(1 mark)**

2. Explain how leaves are adapted for photosynthesis.   **(4 marks)**

3. List three factors that affect the rate of photosynthesis.   **(1 mark)**

Environmental Damage

Changes in the environment affect the distribution of living organisms.

Animals and plants are subjected to environmental changes. Such changes may be caused by living or non-living factors such as a change in a competitor, or in the average temperature or rainfall.

Living organisms can be used as **indicators** of climate change and pollution. For example, lichens can be used as air pollution indicators. Lichens are particularly sensitive to the concentration of sulfur dioxide in the atmosphere.

Invertebrates, such as the mayfly nymph and freshwater shrimps, can be used as water pollution indicators. They can also be used as indicators of the concentration of dissolved oxygen in water.

Measuring Factors

Environmental changes can be measured using non-living indicators such as oxygen, nitrate and carbon dioxide levels, as well as temperature and rainfall. Specialist equipment can be used to measure these environmental factors. For example, a light meter can measure light intensity.

Pollution

The human population on Earth has been increasing for a long time, but is now growing more rapidly than ever. Birth rates are increasing and death rates are decreasing.

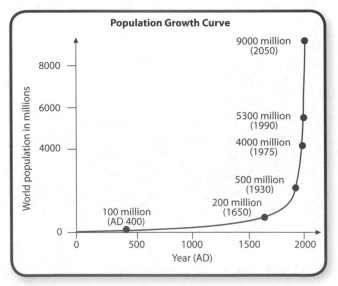

The rate of increase of the population is increasing and this is called **exponential growth**. Combined with an increase in standards of living, increasingly more waste is being produced. Unless waste is properly handled, more pollution will be caused.

Waste may pollute:

- water with sewage, fertilisers or toxic chemicals
- air with smoke and gases such as sulfur dioxide
- land, with toxic chemicals such as pesticide and herbicides that may be washed from the land into waterways.

A pollutant is a substance that harms living things. **Burning fossil fuels** is the main cause of atmospheric pollution and it releases **carbon dioxide** that contributes to the **greenhouse effect**. It also releases **sulfur dioxide** and **nitrogen oxides** that cause **acid rain**. **Methane** gas from cattle and rice fields is another greenhouse gas.

Humans reduce the amount of land available for other animals and plants by building, quarrying, farming and dumping waste.

Fertilisers

Plants need nutrients to grow, which they take up from the soil. **Intensive farming** uses up nutrients quickly, so the farmer has to replace them with **artificial fertilisers**.

Fertilisers enable farmers to produce more crops from a smaller area of land. This can reduce the need to destroy the countryside for extra agricultural space. There is a problem, however, when fertilisers wash into rivers and lakes. This is called **eutrophication**.

Eutrophication is when untreated sewage and fertilisers may cause surface plants and algae to grow rapidly. This blocks out sunlight to plants below, restricting their photosynthesis.

When the plants and algae die, the microbes, which break them down, increase in number. The microbes use up the dissolved oxygen in the water. Animals that live in the water, including fish, may suffocate and die.

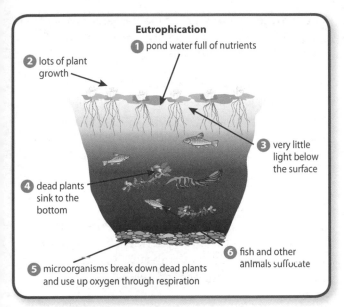

Eutrophication

1 pond water full of nutrients

2 lots of plant growth

3 very little light below the surface

4 dead plants sink to the bottom

5 microorganisms break down dead plants and use up oxygen through respiration

6 fish and other animals suffocate

Deforestation

Deforestation may take place in order to…

● provide timber
● allow the growth of crops (e.g. for biofuels)
● create land for cattle and rice fields in order to produce more food (to try and provide for the growing population).

Deforestation causes several problems:

● It increases the release of carbon dioxide into the air due to burning timber and activities of microorganisms.
● It reduces the rate at which oxygen is released and carbon dioxide is removed from the atmosphere and 'locked up' for many years as wood.
● It leads to soil erosion as the soil is exposed to rain and wind.
● Water from trees evaporates into the air and without it there will be a decrease in rainfall.
● Destroying forests destroys many different habitats for animals and plants, reducing biodiversity.

Destruction of Peat

The destruction of peat bogs and other areas of peat releases carbon dioxide into the atmosphere. Using peat-free composts could help to reduce this problem.

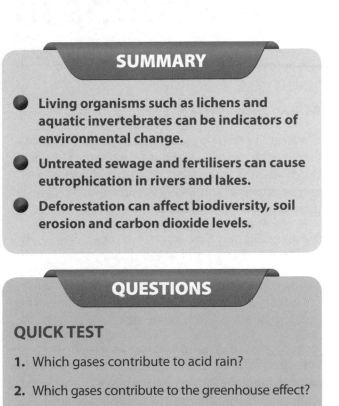

SUMMARY

● **Living organisms such as lichens and aquatic invertebrates can be indicators of environmental change.**

● **Untreated sewage and fertilisers can cause eutrophication in rivers and lakes.**

● **Deforestation can affect biodiversity, soil erosion and carbon dioxide levels.**

QUESTIONS

QUICK TEST

1. Which gases contribute to acid rain?

2. Which gases contribute to the greenhouse effect?

3. What type of species might be used as an indicator for identifying polluted water?

EXAM PRACTICE

1. Describe and explain the stages of eutrophication that lead to decreased biodiversity in lakes.

   The quality of your written communication will be assessed in your answer to this question.

   (6 marks)

2. Many forests on the planet have been cut down by humans.

   a) What is deforestation? **(1 mark)**

   b) Give two reasons why forests are removed. **(2 marks)**

   c) Describe two problems that deforestation has on wildlife. **(2 marks)**

3. Explain why preserving forests would help reduce the greenhouse effect. **(2 marks)**

Interactions in Environments

The world is inhabited by millions of different plant and animal species, with probably many more yet to be discovered.

Maintaining Biodiversity

Biodiversity refers not just to the numbers of species, but also the genetic variations within those species and the differences in the many different habitats on the Earth. Reducing biodiversity would remove variation that may be vital for future survival of species.

Biodiversity provides us with many benefits, such as food, medicines, air and water.

Taking more than we need from the environment, for example, chopping down trees and not replacing them, or hunting, is not sustainable.

More people are using up more resources with more intensity than at any point in human history. **Sustainable development** means meeting the needs of people today, without damaging the Earth for future generations.

Maintaining biodiversity to ensure the **conservation** of different species is one of the keys to sustainability. Large-scale monoculture crop production is not sustainable because it does not maintain biodiversity.

People around the world can have different impacts on the environment depending on their economic and industrial conditions. Sustainability can be improved by reducing, reusing and recycling. For example, reducing the amount of packaging, or using biodegradable products.

Extinction

Species that are unable or too slow to adapt to changes in their surroundings become extinct. Examples of extinct species include the mammoth, the dodo and the sabretooth tiger.

Extinction can also be caused by changes in the environment, new predators, new disease, new successful competition, a single catastrophic event (e.g. volcanic eruption), the cyclical nature of speciation, or human activity, e.g. hunting, pollution or habitat destruction.

An **endangered species** is a plant or animal that is in danger of becoming extinct, e.g. the panda, gorilla, whale, red squirrel and osprey.

We need to look at ways of protecting habitats, introducing quotas, breeding in captivity, giving endangered species legal protection, education programmes, seed banks, and even creating artificial ecosystems in which to keep and breed animals.

Interdependence

Organisms in an environment interact and rely on each other for life. We say they are **interdependent**. A **food web** shows the interdependence of organisms with each other. Factors such as over-fishing of North Sea cod will affect aquatic food webs.

Animals and plants adapt to where they live in order to survive. They compete with each other for resources and are under threat from predators and disease. These factors and interactions between animals, plants and their environment affect population numbers and distribution in a community.

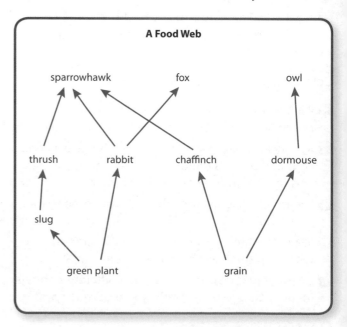

A Food Web

Some organisms benefit from the presence of organisms from a different species – this is **mutualism**. For example, the oxpecker is a bird that eats insects on the backs of buffalos, and remora fish clean sharks.

Extreme Environments

In extreme environments, the conditions determine which animals and plants exist and survive.

Deep-sea volcanic vents were only discovered in 1977. Occurring along the mid-ocean ridge, they are regions on the ocean floor where fluids and gases such as methane and hydrogen sulfide seep through from the Earth's crust. These fluids and gases are so deep in the ocean that there is no light for photosynthetic organisms to survive.

Although extreme, food webs do exist. The primary producers are bacteria that use energy from hydrogen sulfide and oxygen to make food, which is then used by other organisms. Because they are using chemicals and not light to obtain energy, the process is called **chemosynthesis**.

Chemosynthetic bacteria in tube worms in deep-sea vents are an example of mutualism. Proteins in the bacteria are able to withstand the extremely high temperatures (up to 400°C); they are resistant to heat.

Other extreme environments that have promoted adaptations to survival include the Antarctic (where animals have had to adapt to the extreme cold), and high altitudes (where there is a lack of oxygen).

Soil

Soil is important for plant growth. Soils are a mixture of different-sized mineral particles, dead material (**humus**), living organisms (e.g. worms), air and water.

Sandy soils drain better than clay soils that can become waterlogged. Loam is a soil that contains a mixture of clay and sand. It might be necessary for farmers or gardeners to neutralise acidic soils by adding lime or chalk.

Some living organisms in soil depend on a supply of oxygen and water. Humus is important because it releases minerals when it decomposes and increases the air content of soil.

Earthworms are important to soil structure and fertility. Earthworms…

- bury organic material for decomposition by bacteria and fungi
- aerate and drain the soil
- mix up soil layers and neutralise acidic soils.

QUESTIONS

QUICK TEST

1. What is sustainable development?
2. What is biodiversity?
3. What does the term interdependence mean?

EXAM PRACTICE

1. Deep-sea volcanic vents are regions of ocean floor where there is no light and only gases such as methane and hydrogen sulfide.

   a) Describe how organisms such as bacteria survive in these vents. **(2 marks)**

   b) Name two other extreme environments. **(2 marks)**

   c) What is the difference between photosynthesis and chemosynthesis? **(2 marks)**

Ecology and Classification 1

We are surrounded by a huge variety of living organisms in a variety of habitats. In any habitat, the number of organisms is usually large, so we use sampling techniques to find out what lives there and then estimate the size of the population.

There are a variety of techniques used to sample organisms.

Quadrats

A **quadrat** is a wooden or metal frame, usually 1 m² in area. It is placed randomly in an area, e.g. a field, and the number of plants belonging to a species is counted within the quadrat. This is repeated to estimate the number of plants in the whole area.

Transects are sometimes used to show the abundance and distribution of a species across an environmental gradient, for example, the distribution of plant species from a shaded area into a light area.

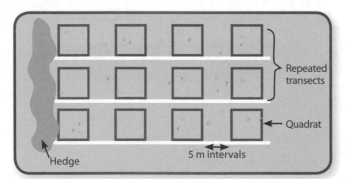

Repeated transects

Quadrat

5 m intervals

Hedge

Zonation is a gradual change of an abiotic factor (e.g. light intensity or gradient of slope) that can result in the zonation of organisms in a habitat. For example, different types of organisms live at different zones in a rocky shore.

Sampling Animal Populations

Insects tend to move around, so **pooters** and **pitfall traps** can be used to collect them. The insects can then be counted and identified.

When sampling mobile populations, such as beetles in a woodland, scientists use a technique where they capture the beetles, mark them in such a way as not to affect them behaviourally, and then release them back into their habitat. The scientists then recapture a second sample, count the number of marked individuals and carry out simple calculations to estimate the population size.

Scientists need to make certain assumptions when using **capture–recapture** data, including…

- whether there has been any deaths, immigration or emigration
- that identical sampling methods have been used
- that marking has not affected the survival rate, e.g. by increased predation due to animals being made more identifiable to predators.

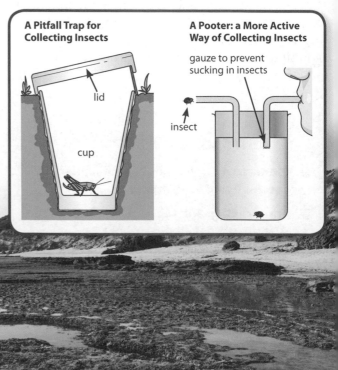

A Pitfall Trap for Collecting Insects

lid

cup

A Pooter: a More Active Way of Collecting Insects

gauze to prevent sucking in insects

insect

Classification

Living organisms show a range of sizes, features and complexity. **Classification** is sorting living things into groups.

Carl Linnaeus devised a classification system based on common features, such as body shape, type of limbs and skeleton. He looked for similarities and differences. Nowadays, scientists can use DNA to help classify organisms more accurately. Classification helps us to understand evolutionary and ecological relationships.

From the largest to the smallest group, the order of classification is: **Kingdom**, **Phylum**, **Class**, **Order**, **Family**, **Genus**, **Species**.

Keys

Keys can be used to identify the organisms collected.

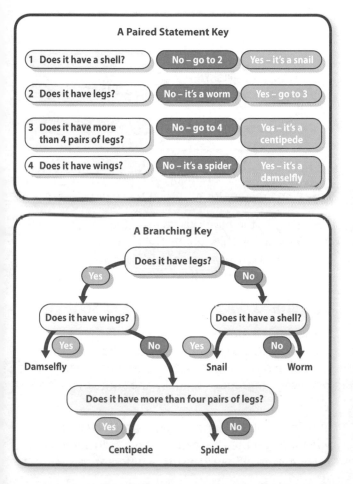

A Paired Statement Key

1 Does it have a shell? — No – go to 2 — Yes – it's a snail

2 Does it have legs? — No – it's a worm — Yes – go to 3

3 Does it have more than 4 pairs of legs? — No – go to 4 — Yes – it's a centipede

4 Does it have wings? — No – it's a spider — Yes – it's a damselfly

A Branching Key

Does it have legs?
Yes / No

Does it have wings?
Yes → Damselfly / No

Does it have a shell?
Yes → Snail / No → Worm

Does it have more than four pairs of legs?
Yes → Centipede / No → Spider

SUMMARY

- Scientists use sampling methods to estimate and monitor populations.

- Quadrats and transects are used to sample plant populations.

- Pooters, pitfall traps and capture–recapture techniques are used to sample mobile populations.

- Keys are used to help scientists correctly identify organisms they have collected.

QUESTIONS

QUICK TEST

1. List the order of classification system, from the largest to the smallest group.

2. Why do scientists use sampling techniques?

3. What assumptions are made when using capture–recapture techniques?

EXAM PRACTICE

1. A group of students investigated the number of different plant species found on a hillside.

   a) What piece of equipment are the students likely to use? **(1 mark)**

   b) Describe how the students would carry out this investigation. **(4 marks)**

   c) One student noticed that different species grew at the top of the slope compared to the bottom of the slope. What sampling technique would the students use to investigate this? **(1 mark)**

   d) Explain why one member of the group suggested that using a key might be useful. **(1 mark)**

Ecology and Classification 2

Kingdoms

Today scientists recognise five **kingdoms** (described in the table below) but recently the 3 domain classification system has been preferred.

Kingdom	Main Characteristics	Picture
Animalia	Multicellular, do not have cell walls, do not have chlorophyll, feed heterotrophically	
Plantae	Multicellular, have cell walls, have chlorophyll, feed autotrophically, divided into non-flowering and flowering plants	
Fungi	Multicellular, have cell walls, do not have chlorophyll, feed saprophytically	
Protoctista	Unicellular, have a nucleus	
Prokaryotes	Unicellular, have no nucleus	

Scientists do not classify **viruses** in any of the five kingdoms and regard them as **non-living**.

Heterotrophic feeders rely on obtaining food from their environment and digesting it.

Autotrophic feeders are able to make their own food using photosynthesis.

Saprophytic feeders are organisms that feed on dead material.

The animal kingdom is divided into **vertebrates** (animals with backbones) and **invertebrates** (animals without backbones).

The vertebrates are known as the phylum **Chordata**, and this is divided into five classes. The features that each group has in common include:

- How they absorb oxygen, e.g. lungs, gills or through the skin.
- How they reproduce, e.g. internal or external fertilisation, **oviparous** (lay eggs) or **viviparous** (do not lay eggs).
- How they regulate body temperature, e.g. **homeotherms** (animals able to regulate body temperature – birds and mammals) and **poikilotherms** (animals that cannot regulate body temperature – reptiles, amphibians and fish).

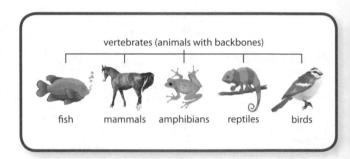

vertebrates (animals with backbones)

fish   mammals   amphibians   reptiles   birds

Some organisms are difficult to assign to one group based on their anatomy and reproduction.

For example, the **platypus** is considered a mammal as it is covered in hair, has mammary glands and is warm blooded. However, the platypus also has some features of a reptile, i.e. it lays eggs and is semi-aquatic.

Species

A **species** is a group of living things that are able to breed together to produce **fertile offspring**. This definition has limitations because some organisms do not always reproduce sexually and some hybrids are fertile.

Individuals of the same species may show great variation, but they have more features in common than they do with organisms of different species.

Binomial classification is where a species is given two scientific names. The first shows the genus it belongs to, and the second shows its species. For example, *Homo sapiens* is the binomial name for humans. Accurate classification means scientists can identify species correctly, which helps when planning conservation efforts.

Classification is complicated by variation within a species, i.e. hybridisation and ring species. Different species can still be very similar and live in similar types of habitat; they may share a common ancestor.

Hybrids occur when closely related species are able to breed together and produce fertile offspring, e.g. Mallard ducks can breed with American Black ducks.

Populations of the same species living close by may have slightly different characteristics, but are still able to interbreed. Sometimes there is a chain of different populations that can all breed with their neighbouring populations, but the two populations at either end of the chain can't interbreed. These organisms are often called **ring species** as the chain can form a ring shape. Examples include different gull species around the Arctic and different salamander species in South America.

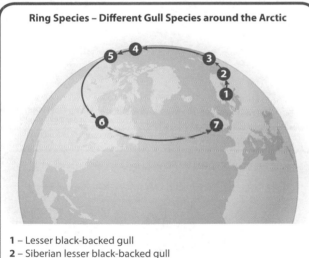

Ring Species – Different Gull Species around the Arctic

1 – Lesser black-backed gull
2 – Siberian lesser black-backed gull
3 – Heuglins gull
4 – Birula's gull
5 – Vega Herring gull
6 – American Herring gull
7 – Herring gull

Gulls interbreed with their geographical neighbours in a continuous ring. Numbers 1 and 7 do not interbreed with each other.

SUMMARY

- **Classification seeks to organise organisms by looking at similarities and differences, and has placed organisms into five kingdoms.**

- **A species is a group of similar organisms that can breed to produce fertile offspring.**

- **Hybrids are the offspring from two different species.**

- **A chain of related species living geographically close are called ring species.**

QUESTIONS

QUICK TEST

1. Define the term species.

2. How many kingdoms exist today?

3. What does 'vertebrate' mean?

EXAM PRACTICE

1. When a lion and a tiger mate, the offspring is called a liger.

   a) Is the offspring likely to be fertile or infertile? Explain your answer. **(2 marks)**

   b) The two-part name for lions is *Panthera leo*. State which genus the lion belongs to. **(1 mark)**

2. Circle the information that is the least likely to be useful in classifying an animal. **(1 mark)**

   habitat   anatomy   fossil record   DNA

3. Explain why plants are called autotrophs. **(1 mark)**

Adaptation and Competition

A **habitat** is where an organism lives and has the conditions needed for it to survive. A **community** consists of living things in the habitat.

An **ecological niche** is the role that an organism takes in an ecosystem.

Each community is made up of different populations of animals and plants, and each population is adapted to live in that particular habitat. An **ecosystem** is formed from all the living things and their physical environment.

There are natural ecosystems such as woodlands or lakes, or **artificial ecosystems**, such as greenhouses and aquariums. Artificial ecosystems tend to have a smaller variety of organisms and may use weedkillers, pesticides and fertilisers to control growth.

Competition

In order to survive and reproduce, organisms require a supply of materials from their surroundings and from the other living organisms there. In a community, the animal or plant that is best adapted to its surroundings will survive and be more able to compete for limited resources.

Plants often compete with each other for light and space, and for water and nutrients from the soil. Animals often compete with each other for food, mates and territory.

Intraspecific competition occurs between individuals of the same species, whereas **interspecific** competition occurs between individuals of different species.

Size of Population

Factors that keep the population from becoming too large are called **limiting factors**.

The factors that affect the size of a population are:
- amount of food and water available
- availability of light, oxygen and carbon dioxide
- predators / grazing
- disease
- climate, temperature, floods, droughts and storms
- competition
- human activity, such as pollution or habitat loss.

Adaptations

Organisms, including microorganisms, have **adaptations** (special features) and behaviours (e.g. migration and hibernation) that help them to survive in the conditions in which they normally live.

Organisms that live in environments that are very extreme are called **extremophiles**. For example, they may be tolerant to high levels of salt, high temperatures or high pressures.

Some animals and plants may be adapted to cope with specific features of their environment, e.g. thorns, poisons, mimicry, living in groups, and warning colours to deter predators.

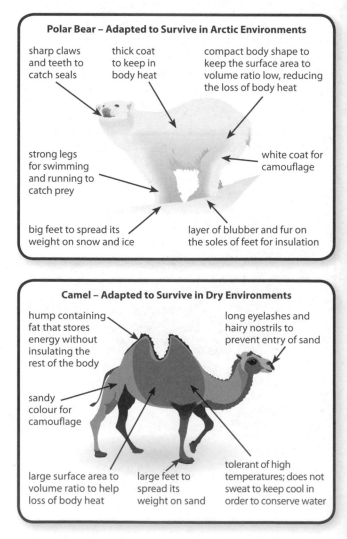

Polar Bear – Adapted to Survive in Arctic Environments

sharp claws and teeth to catch seals

thick coat to keep in body heat

compact body shape to keep the surface area to volume ratio low, reducing the loss of body heat

strong legs for swimming and running to catch prey

white coat for camouflage

big feet to spread its weight on snow and ice

layer of blubber and fur on the soles of feet for insulation

Camel – Adapted to Survive in Dry Environments

hump containing fat that stores energy without insulating the rest of the body

long eyelashes and hairy nostrils to prevent entry of sand

sandy colour for camouflage

large surface area to volume ratio to help loss of body heat

large feet to spread its weight on sand

tolerant of high temperatures; does not sweat to keep cool in order to conserve water

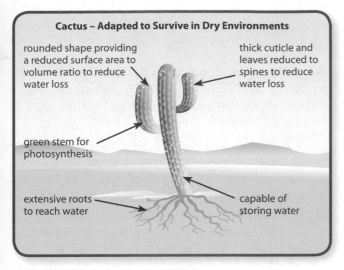

Cactus – Adapted to Survive in Dry Environments

rounded shape providing a reduced surface area to volume ratio to reduce water loss

thick cuticle and leaves reduced to spines to reduce water loss

green stem for photosynthesis

extensive roots to reach water

capable of storing water

Microscopic Life in Water

It can be more advantageous for organisms to live in water. There are no problems of water shortage or dehydration. There is less variation in temperature, and water provides structural support. It is also easy to dispose of waste products.

There are some disadvantages – there is resistance to movement and it can be difficult to regulate water content due to osmosis.

Predator/Prey Graphs

In a community, the number of animals stays fairly constant. This is partly due to the amount of food limiting the size of the populations.

A predator is an animal that hunts and kills another animal. A prey is the hunted animal.

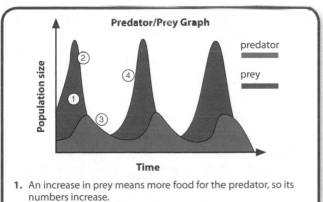

Predator/Prey Graph

Population size — Time

predator

prey

1. An increase in prey means more food for the predator, so its numbers increase.
2. The prey then decreases as it is eaten.
3. Predators then decrease, as there is not enough food.
4. Prey numbers can increase again, and so it continues.

SUMMARY

● Population size is controlled by limiting factors including competition, predation and disease.

● Organisms have adaptations (special features) and different behaviours to help them survive in various habitats.

QUESTIONS

QUICK TEST

1. What is a habitat?

2. Why do the numbers of animals in a community stay fairly constant?

3. What is intraspecific competition?

EXAM PRACTICE

1. Arctic foxes live in cold conditions whereas the Fennec fox lives in desert conditions. Compare and explain how these animals have both adapted to their environment to increase their chance of survival.

Arctic Fox

Fennec Fox

The quality of your written communication will be assessed in your answer to this question.

(6 marks)

2. a) What is meant by the term 'artificial ecosystem'? **(1 mark)**

   b) Why might an 'artificial' ecosystem have less biodiversity than a 'natural' ecosystem? **(2 marks)**

Evolution

The theory of evolution states that all species of living things have evolved from simple life forms that first developed more than three billion years ago.

Religious theories are based on the idea of a 'creator' of all life on Earth, but there are other theories.

In 1859, a British naturalist called Charles **Darwin** put forward his ideas on **evolution**. Darwin had visited the Galapagos Islands off the coast of South America and combined with his research in Kent, England, he made a number of observations that led to his theory:

- Organisms produce more offspring than can possibly survive.
- Population numbers remain fairly constant despite this.
- All organisms in a species show variation.
- Some of these variations are inherited.

Darwin concluded from these observations that, since there were more offspring than could survive, there must be a struggle for existence, competition for food, predators and disease. This led to the strongest and fittest offspring surviving and passing on their genes. This is sometimes called **survival of the fittest** or **natural selection**.

Variation that is inherited forms the basis of evolution. Over a period of time the proportion of individuals with the advantageous characteristics in the population will increase compared with the proportion of individuals with poorly adapted characteristics, and the poorly adapted characteristics may eventually be lost.

The theory of evolution by natural selection was not accepted at first due to several reasons:
- The theory challenged the idea that God made all the animals and plants that live on Earth.
- There was **insufficient evidence** to convince scientists at the time that the theory was published.
- The **mechanisms of inheritance** and variation were not known until 50 years after the theory was published.

Jean-Baptiste **Lamarck** suggested another theory, where changes that occur in an organism during its lifetime can be inherited. For example, he suggested that giraffes grew longer necks because they needed to reach food. We now know that in the vast majority of cases this type of inheritance cannot occur.

Fossils

Fossils are the preserved remains of dead organisms that lived millions of years ago, found in rocks. Fossils can be formed from the hard parts of animals that do not decay easily, or from organisms that have not decayed because one or more of the conditions needed for decay are absent. Fossils can also be formed when parts of the organism are replaced by other materials as they decay, or as preserved traces such as footprints, burrows and rootlet traces.

Fossils provide evidence for evolution. However, there are gaps in the fossil record because fossils do not always form, soft tissue decays and many fossils have not yet been found. Also, many fossils have been destroyed by geological activity. Work continues in this field to fully understand the process of evolution as further evidence from genetics and molecular biology is collected.

The **pentadactyl limb** (5-digit hand and foot) provides evidence for evolution. Many animals, including mammals, birds, dinosaurs and other reptiles and amphibians, all have pentadactyl limbs. This suggests that there is a **common ancestor** for all these animals.

Scientists now believe that humans did not evolve from apes, but that humans and apes share a common ancestor that lived millions of years ago. Evidence for human evolution is based on fossils and **stone tools**. Tools can be dated from their environment.

- **Ardi** is a fossil from 4.4 million years ago found in Ethiopia, and is the oldest partial skeleton from a human ancestor to be discovered.
- **Lucy**, another fossil from Ethiopia, is dated at 3.2 million years ago. She has features like a chimpanzee (small brain, short legs and long arms) but also has features in her knee and pelvis that suggests she walked upright like humans.

Since the 1930s the **Leakey** family have been investigating the origins of humankind in Africa by discovering fossils from 1.6 million years ago.

Evolution in Action

Evolution is an ongoing process that scientists see today. Examples are the evolution of bacteria that are resistant to antimicrobial chemicals, or rats that are resistant to the pest control product, Warfarin.

An example of the environment causing changes in a species is that of the **peppered moth**. The moths live in woodlands on lichen-covered trees. There are two types of peppered moth: a light, speckled form and a dark form. The dark form was caused by a **mutation** and tended to be eaten by predators.

Dark Peppered Moth

Peppered Moth

In the 1850s, the dark type of moth was rare, but pollution from factories started to blacken tree trunks. The dark moth was then at an advantage because it was **camouflaged**.

In 1895, most of the population of moths were dark. In cleaner, less polluted areas, the light moth had an advantage against predators and so it still survived to breed.

New Species

New species arise as a result of **isolation** where two populations of a species become separated (for example, by river or mountain range).

The new species can arise as a result of several factors:

- **Genetic variation** – each population has a wide range of alleles that control their characteristics.
- **Natural selection** – within each population, alleles are selected that control the characteristics that help the organism to survive.
- **Speciation** – the populations become so different that successful interbreeding is no longer possible.

SUMMARY

- **Darwin based his theory of evolution through natural selection.**
- **Fossils help to provide evidence for human evolution, including fossils such as Ardi and Lucy.**
- **Extinction occurs when organisms are unable or too slow to adapt to changes in their environment.**
- **New species arise due to a combination of isolation, natural selection, genetic variation and speciation.**

QUESTIONS

QUICK TEST

1. What does 'survival of the fittest' mean?
2. Describe Lamarck's theory of evolution.
3. What are fossils?

EXAM PRACTICE

1. Give two reasons why Darwin's theory was not accepted by some people. **(2 marks)**
2. Describe Darwin's theory of evolution by natural selection.

   The quality of your written communication will be assessed in your answer to this question.

   (6 marks)

Cells

Cells are the building blocks of life. All living things are made up of cells.

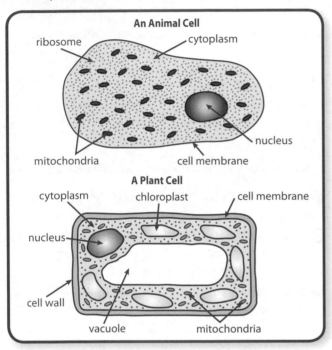

An Animal Cell

ribosome, cytoplasm, nucleus, cell membrane, mitochondria

A Plant Cell

cytoplasm, chloroplast, cell membrane, nucleus, cell wall, vacuole, mitochondria

Most human and animal cells contain the following:

- A **nucleus**, which contains genetic information and controls the chemical reactions taking place inside the cell.
- The **cytoplasm**, where most chemical reactions, controlled by enzymes, take place.
- The **cell membrane**, which controls the passage of substances in and out of the cell.
- The **mitochondria** where most energy is released by respiration.
- **Ribosomes** in the cytoplasm, which are the site of protein synthesis.

Plant and algal cells also contain:

- A **cell wall** made of cellulose, which gives a plant cell strength and support.
- A **vacuole** containing a weak solution of salt and sugar called cell sap; this also gives cell support.
- **Chloroplasts** containing chlorophyll to absorb the Sun's energy, and enzymes for the reactions in photosynthesis.

Microscopes allow us to study cells in detail. Many structures in a cell can be seen with a light microscope (except ribosomes), but the electron microscope has a far greater magnifying power because it uses electrons instead of light to illuminate the object. The formula used to calculate magnification is:

$$\text{Magnification} = \frac{\text{image size}}{\text{actual size}}$$

Special Cells

Cells are different shapes in order to carry out particular functions efficiently.

Specialised Animal Cells

A **sperm** cell has a tail that enables it to swim towards the egg. Its head is also streamlined to aid swimming and contains enzymes to penetrate the egg. Sperm cells are produced in large numbers to increase the chance of fertilisation.

The **egg cell** contains nutrients in the cytoplasm. Immediately after fertilisation the cell membrane around the egg changes to block other sperm from entering.

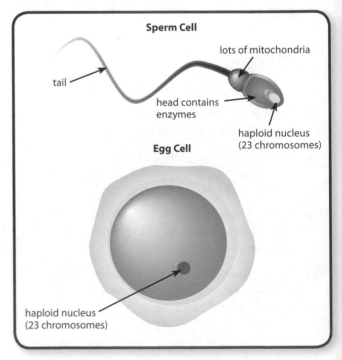

Sperm Cell

tail, lots of mitochondria, head contains enzymes, haploid nucleus (23 chromosomes)

Egg Cell

haploid nucleus (23 chromosomes)

Specialised Plant Cells

Root hair cells are long and thin, providing the plant with a large surface area for absorption. Water is taken up by osmosis and mineral salts are taken up by **active transport**.

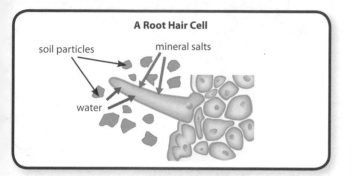

A Root Hair Cell

soil particles
mineral salts
water

Palisade cells have **many chloroplasts**. These cells are near the upper surface of the leaf so the chloroplasts can absorb sunlight for photosynthesis.

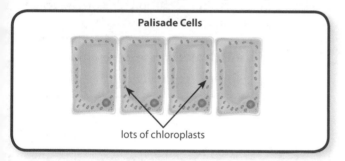

Palisade Cells

lots of chloroplasts

Stem Cell Therapy

Stem cells are cells that have the ability to replicate and specialise into different types of tissue throughout the life of the organism. They are found in adult bone marrow, the brain, blood and heart, human embryos and the umbilical cord. Up to the eight-cell stage, **embryonic stem cells** have the ability to divide and specialise into any tissue needed, such as nerve cells. After the eight-cell stage, most of the embryo cells become specialised and form different types of tissue.

Stem cells from adults **do not** have the same ability to change into any type of cell; they can only change into a cell from where they originated. These stem cells are also scarce in the body and are harder to culture.

The potential for stem cell therapy is great. It can be used to replace tissue that has lost its function, such as damaged heart tissue, and used to treat genetic diseases.

In the future, stem cells could be used to treat burn victims, spinal cord injuries, heart disease, cancer (including the treatment of leukaemia) and diabetes. It could mean the end of mechanical devices such as artificial joints and plastic arteries by using living, natural replacements.

The use of stem cells is very **controversial**. Human embryos are destroyed during the process. Some people believe this is wrong as the embryo is a potential life and that it is murder. Others believe that using stem cells develops our knowledge of science and can benefit everyone.

SUMMARY

- The different parts of cells carry out different functions.
- Some cells are specialised to carry out specific roles.
- Stem cells are able to specialise into different types of cells and could be used to treat various patients.

QUESTIONS

QUICK TEST

1. What occurs in the mitochondria?

2. How are root hair cells adapted to carry out their function?

3. Define the term 'stem cell'.

EXAM PRACTICE

1. List the structures that plant cells have and animal cells do not. **(3 marks)**

2. Stem cell therapy is very controversial. Explain the different sides of the argument.

   The quality of your written communication will be assessed in your answer to this question.

   (6 marks)

Diffusion and Osmosis

Dissolved substances can move into and out of cells by diffusion, osmosis and active transport. As the size and complexity of organisms increases it becomes more difficult to exchange materials, so special adaptations are needed.

Simple Diffusion

Diffusion is the movement of particles from **an area of high concentration to an area of low concentration** until they are evenly spread out.

Rules to Remember

Diffusion occurs at a faster rate if:

- the particle size is smaller
- the diffusion distance is smaller
- the surface area for diffusion is increased
- the difference in concentration is large (this is called the **concentration gradient**).

Diffusion in Animal Cells

Only certain substances can pass through the cell membrane by diffusion. Diffusion does not require energy. Glucose diffuses into the blood from the small intestine and oxygen diffuses into the blood from the alveoli in the lungs. They are carried in the blood to the cells where they diffuse into the respiring cells.

Carbon dioxide diffuses from respiring cells into capillaries and travels to the lungs. Carbon dioxide and oxygen are exchanged in the alveoli of the lungs by diffusion.

Many organ systems are specialised for exchanging materials, e.g. the digestive system in animals.

The inner surface of the small intestine contains many **villi** that increase the surface area for absorption of food molecules by diffusion into the blood. The villi also have a good blood supply to maintain a concentration gradient.

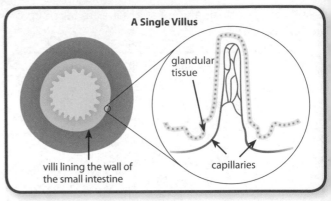

A Single Villus

glandular tissue

villi lining the wall of the small intestine

capillaries

Diffusion in Plant Cells

During photosynthesis, carbon dioxide diffuses into the leaf via the stomata found on the underside of the leaf. **Leaves** are thin so carbon dioxide can reach the inner cells easily and have plenty of stomata in the lower surface for gas exchange by diffusion. Oxygen and water vapour diffuse out of the **stomata**, particularly in hot, dry, windy conditions.

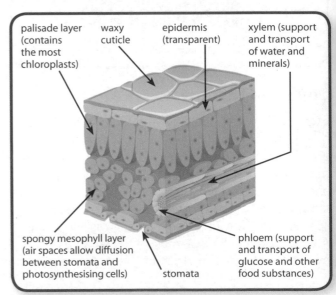

palisade layer (contains the most chloroplasts)

waxy cuticle

epidermis (transparent)

xylem (support and transport of water and minerals)

spongy mesophyll layer (air spaces allow diffusion between stomata and photosynthesising cells)

stomata

phloem (support and transport of glucose and other food substances)

Osmosis

Osmosis is the movement of **water molecules** from an area of **high water concentration** (weak/dilute solution) to an area of **low water concentration** (strong/concentrated solution) through a **partially permeable membrane**.

'Partially permeable' means it allows small molecules to pass through but not larger ones.

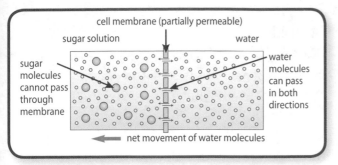

Osmosis in Plant Cells

Root hairs take in water from the soil by osmosis. Water continues to move along the cells of the root and up the xylem to the leaf.

Osmosis makes plant cells swell up. The water moves into the plant cell vacuole and pushes against the cell wall. The cell wall stops the cell from bursting. We say that the cell is **turgid**. This is useful as it gives plant stems support.

If a plant is lacking in water then it wilts and the cells become **flaccid** as water moves out of the cells. If a lot of water leaves the cells then the cell membrane starts to peel away from the cell walls. The cells have undergone **plasmolysis**.

Osmosis in Animal Cells

Animal cells have no cell wall to stop them swelling (**crenation**). So if they are placed in pure water they take in water by osmosis until they burst – this is called **lysis**.

Active Transport

Active transport happens where substances enter cells **against a concentration gradient**. This requires energy from respiration. Active transport allows cells to absorb ions from very dilute solutions.

SUMMARY

- **Dissolved particles move into and out of cells by diffusion, osmosis and active transport.**

- **Plants and animals have adapted to increase the rate of diffusion.**

QUESTIONS

QUICK TEST

1. Define diffusion.

2. What is a partially permeable membrane?

3. Why do animal cells burst if they are placed in pure water?

EXAM PRACTICE

1. Describe and explain how the features of alveoli in the lungs make them efficient at diffusion. **(4 marks)**

2. For each of the following substances, identify if it moves between the cells and blood by osmosis or diffusion.

   a) Glucose   b) Oxygen

   c) Urea   d) Water

   e) Carbon dioxide **(5 marks)**

3. Dylan and Molly investigated the effect of four different salt solution concentrations on the mass of potato chips. They weighed each chip and then placed one into each solution. After an hour they removed the chips and reweighed them. Their results are given in the table.

	Conc. of salt solution			
	0M	1M	2M	3M
Mass (g) of chip at start	2.6	2.5	2.6	2.7
Mass (g) of chip after 1hr	2.8	2.3	2.1	1.9

   a) In which concentration of salt solution did the chip gain mass? **(1 mark)**

   b) Explain why the chip gained mass. **(2 marks)**

   c) Molly doesn't think that their results are reliable. What should they do next time? **(1 mark)**

   d) Dylan doesn't think their results are accurate. What should they do next time? **(1 mark)**

Meiosis

Reproductive Systems

The male reproductive system includes the testes contained in a scrotum sac outside of the body because this is a more suitable temperature for sperm development. The testes produce the hormone **testosterone** as well as sperm that are carried to the penis via sperm ducts.

The female reproductive system includes two ovaries that produce eggs. An egg is released once a month into the oviduct, where fertilisation can take place. The fertilised egg moves to the uterus where it implants into the wall and grows into a foetus.

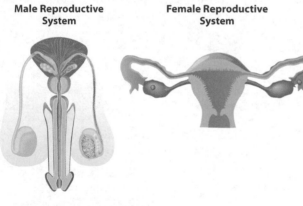

Male Reproductive System

Female Reproductive System

Mitochondrial DNA

When an egg is fertilised, only the nucleus from the sperm enters the egg. All of the **mitochondrial DNA** comes from the mother, not the father. Unlike nuclear DNA it does not recombine, but follows a direct line of descent unchanged from mother to offspring.

Mitochondrial DNA is less likely to break down over time than nuclear DNA. It can be extracted from fossils.

Mitochondrial DNA mutates about 20 times faster than nuclear DNA. These mutations are key for tracing human ancestry and migration. Each mutation is a genetic marker. As scientists can date fossils, they can also date the markers.

Scientists have traced the markers and have learned about migration patterns in our human ancestors. They have identified our most recent *Homo sapiens* common ancestor – **African Eve**.

Meiosis

Meiosis is a type of cell division that occurs in the formation of **gametes** (sex cells). It occurs in cells in the reproductive organs (testes and ovaries in humans).

Meiosis produces cells that have half the number of chromosomes of body cells where they are normally found in pairs. They are called **haploid cells** and are the gametes. In humans the haploid number is 23. Fertilisation restores the normal number of chromosomes to 46 (**diploid cells**) which are the body cells.

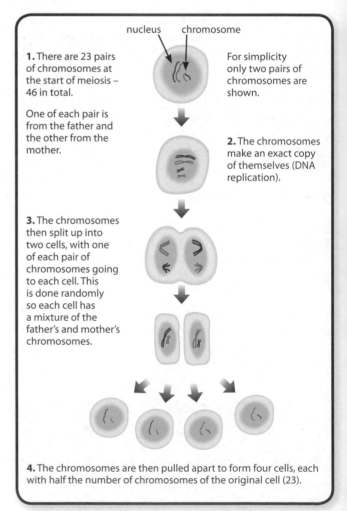

nucleus   chromosome

1. There are 23 pairs of chromosomes at the start of meiosis – 46 in total.

One of each pair is from the father and the other from the mother.

For simplicity only two pairs of chromosomes are shown.

2. The chromosomes make an exact copy of themselves (DNA replication).

3. The chromosomes then split up into two cells, with one of each pair of chromosomes going to each cell. This is done randomly so each cell has a mixture of the father's and mother's chromosomes.

4. The chromosomes are then pulled apart to form four cells, each with half the number of chromosomes of the original cell (23).

Meiosis also occurs in plants in the formation of pollen and ovules.

The Inheritance of Sex

When all the chromosomes are sorted into pairs, the **23rd** pair are the **sex chromosomes**. These determine whether an individual is a boy or a girl. All the other chromosomes contain information for an individual's characteristics.

A male will have one Y chromosome and one X chromosome. A female will have two X chromosomes. The female ovary will produce only X chromosome eggs during meiosis. The male testis will produce half X chromosome sperm and half Y chromosome sperm. During fertilisation, the egg may join with either an X sperm or a Y sperm.

A **mutation** is a change in the chemical structure of a gene or chromosome, which alters the way an organism develops. Mutations can be random or there can be a cause. Mutation rates can be increased by ionising radiation. Mutations can be harmful, beneficial or neither. Mutations that occur in reproductive cells are inherited.

The sex-determining gene on the Y chromosome triggers the development of testes in the male reproductive system. The absence of the Y chromosome allows the ovaries to develop in the female reproductive system.

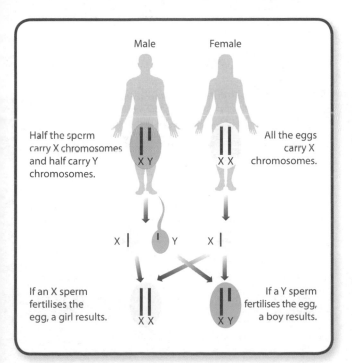

Male   Female

Half the sperm carry X chromosomes and half carry Y chromosomes.

X Y   X X

All the eggs carry X chromosomes.

X | | Y   X |

If an X sperm fertilises the egg, a girl results.

X X

If a Y sperm fertilises the egg, a boy results.

X Y

SUMMARY

- Cell division by meiosis forms gametes with half the genetic information of body cells.

- Males have XY chromosomes and females XX chromosomes.

QUESTIONS

QUICK TEST

1. What does diploid mean?

2. What causes a change in the chemical structure of a gene or chromosome?

3. In humans, where does fertilisation normally take place?

EXAM PRACTICE

1. Put the following sentences into the correct order, showing the sequence of events in meiosis. **(1 mark)**

   A – Paired chromosomes separate

   B – Two cells are formed

   C – To form a total of four new haploid cells

   D – DNA becomes organised into chromosome pairs

   E – Each double-stranded chromosome now separates

   F – DNA is copied

2. Explain how the chromosome number is restored to 46 after meiosis. **(2 marks)**

Mitosis

Mitosis occurs in **growth**, **replacement** and **repair** of cells, as well as asexual reproduction. Mitosis produces all cells except the sex cells, which are formed by meiosis.

Body cells divide by **mitosis**. Body cells have two sets of chromosomes whereas gametes have only one set. The chromosomes contain the genetic information.

During mitosis copies of the genetic material are made. The cell then divides once to produce two daughter cells that are genetically identical to the original parent cell.

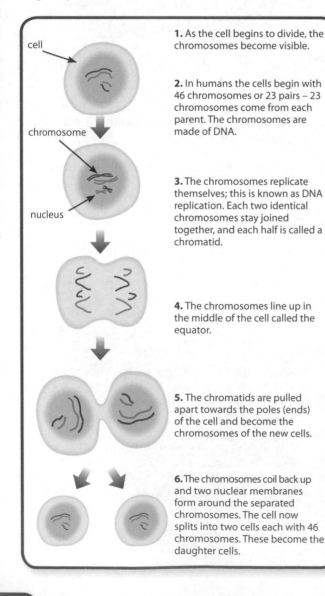

1. As the cell begins to divide, the chromosomes become visible.

2. In humans the cells begin with 46 chromosomes or 23 pairs – 23 chromosomes come from each parent. The chromosomes are made of DNA.

3. The chromosomes replicate themselves; this is known as DNA replication. Each two identical chromosomes stay joined together, and each half is called a chromatid.

4. The chromosomes line up in the middle of the cell called the equator.

5. The chromatids are pulled apart towards the poles (ends) of the cell and become the chromosomes of the new cells.

6. The chromosomes coil back up and two nuclear membranes form around the separated chromosomes. The cell now splits into two cells each with 46 chromosomes. These become the daughter cells.

DNA Replication

The chromosomes are made up of long strands of deoxyribonucleic acid (DNA). **DNA** has the ability to copy itself exactly.

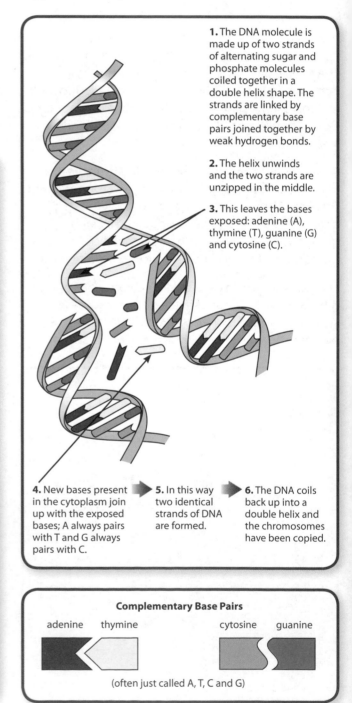

1. The DNA molecule is made up of two strands of alternating sugar and phosphate molecules coiled together in a double helix shape. The strands are linked by complementary base pairs joined together by weak hydrogen bonds.

2. The helix unwinds and the two strands are unzipped in the middle.

3. This leaves the bases exposed: adenine (A), thymine (T), guanine (G) and cytosine (C).

4. New bases present in the cytoplasm join up with the exposed bases; A always pairs with T and G always pairs with C.

5. In this way two identical strands of DNA are formed.

6. The DNA coils back up into a double helix and the chromosomes have been copied.

Complementary Base Pairs

adenine   thymine   cytosine   guanine

(often just called A, T, C and G)

During the 1950s Francis **Crick**, Rosalind **Franklin**, James **Watson** and Maurice **Wilkins** all played roles in the discovery of the structure of DNA.

Watson and Crick used data from other scientists to build a model of DNA. It included X-ray data showing that there were two chains wound in a helix, and other data indicating that bases occurred in pairs.

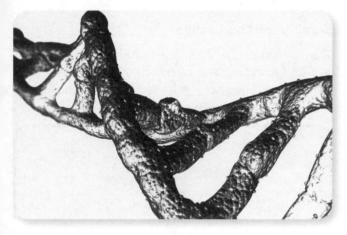

New discoveries, such as Watson's and Crick's, are not accepted immediately. It is important that other scientists can repeat and test their work.

What Happens Next?

When gametes join at **fertilisation**, a single body cell with new pairs of chromosomes is formed. A **zygote** (fertilised egg) starts life with 46 chromosomes. It divides repeatedly by mitosis to form new cells during growth. In multicellular organisms (organisms made up of more than one cell) the cells then differentiate into the different types of cell required.

Although the cells contain identical genes, they use different combinations of them according to the location in the body. Animal cells lose this ability as the animal gets older – the cells in different parts of the body can then only divide to repair and replace themselves.

Mitosis also occurs in plants during growth and replacement. Asexual reproduction occurs through mitosis as it involves only one parent cell. An example includes strawberry plants forming runners.

SUMMARY

- Cell division that results in two identical daughter cells is called mitosis.
- DNA is a double helix.
- There are four bases in DNA: A pairs with T and C pairs with G.

QUESTIONS

QUICK TEST

1. How many daughter cells does mitosis produce?

2. How many chromosomes are in the daughter cells produced by mitosis in humans?

3. What are chromosomes made of?

EXAM PRACTICE

1. State whether the following statements are referring to mitosis or meiosis.　**(3 marks)**

   a) Cell division that produces gametes

   b) Produces cells that are identical

   c) Produces cells that show variation

2. State whether the following statements are true or false.　**(3 marks)**

   a) Mitosis produces genetically identical cells.

   b) In humans, cells produced by mitosis contain 23 chromosomes.

   c) In humans, DNA contains 6 kinds of bases.

Cycles

Substances like carbon, nitrogen and water cycle naturally within the environment.

Nutrients are released during decay, e.g. nitrates and phosphates. These nutrients are taken up by other organisms for growth and other processes, but are eventually returned to the environment as waste or when the organism dies and decays. This results in **nutrient cycles**. In a stable community the processes that remove materials are balanced by processes that return materials.

The Carbon Cycle

Carbon dioxide is a rare atmospheric gas; it makes up approximately **0.04%** of the atmosphere. The amount of carbon released into the atmosphere balances the amount absorbed by green plants and algae.

Photosynthesis

Plants absorb carbon dioxide from the air. They use the carbon to make carbohydrates, fats and proteins that make up the body of plants and algae.

Feeding

Animals eat plants and take the carbon into their bodies to make up their carbohydrates, fats and proteins. Predators eat these animals and so the carbon compounds pass along a food chain.

Respiration

When animals, decomposers, green plants and algae respire, carbon dioxide is released back into the atmosphere.

Death and Decay

Plants and animals die and produce waste. The carbon is released into the soil.

Decomposers

Bacteria and fungi (also known as saprophytes) present in the soil are decomposers that break down dead matter, urine and faeces that contain carbon. Decomposers release carbon dioxide into the atmosphere.

The ideal conditions for decomposition are **warmth**, **moisture** and **oxygen** (aerobic). Without these factors, decay and decomposition cannot take place.

Death, But No Decay

Some plants and animals that die do not decay. Heat and pressure gradually, over millions of years, produce fossil fuels containing carbon.

Fossil Fuels

Coal is formed from plants. Crude oil and natural gas are produced from animals.

Burning and Combustion

The burning of fossil fuels (coal, crude oil and natural gas) and wood releases carbon dioxide into the atmosphere. The energy released from burning fossil fuels originated from the Sun.

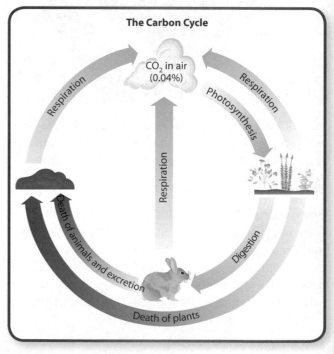

The Carbon Cycle

The Nitrogen Cycle

The atmosphere contains 78% nitrogen gas. **Nitrogen** is an important element needed for making protein. Plants and animals cannot use nitrogen gas. It has to be converted to **nitrates** before plants can use it to make protein.

⬤ Nitrogen is in the air.

⬤ **Lightning** causes nitrogen and oxygen to combine to form nitrogen oxides. These dissolve in rain and are washed into the soil to form **nitrates** in the soil.

- Fertilisers can be added to the soil to improve the nitrate content.
- **Nitrogen-fixing bacteria** in the soil and roots of some plants convert nitrogen from the air into nitrates. They are an example of mutualism.
- Plants take up the nitrates from the soil and convert them into proteins for growth.
- Animals eat the plants and incorporate the protein into their bodies, passing the nitrogen compounds along a food chain or web.
- Animals and plants produce waste; they eventually die and their bodies decay. **Decomposers** such as fungi and bacteria break down this waste and turn the material into ammonium compounds that contain nitrogen.
- **Nitrifying bacteria** in the soil change **ammonia** into nitrates.
- Nitrates can be washed out of the soil before plants take them up. This is called **leaching** and can have serious consequences for rivers and streams.

Denitrifying bacteria live in waterlogged soils. They can change nitrates back into ammonia and nitrogen gas that is returned to the atmosphere.

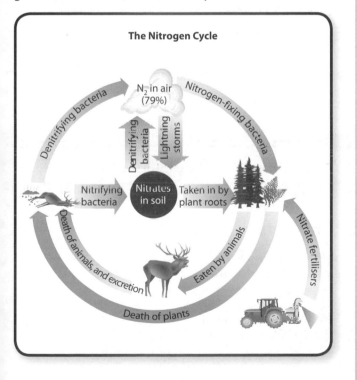

The Nitrogen Cycle

SUMMARY

- **Warmth, moisture and oxygen are needed for decomposition.**
- **Carbon dioxide is removed from the atmosphere by photosynthesis and returned by respiration and combustion.**
- **The nitrogen cycle involves nitrifying bacteria changing ammonia into nitrates.**

QUESTIONS

QUICK TEST

1. Name the process that absorbs carbon dioxide from the air.

2. List two ways in which carbon is released back into the air.

3. What do plants use nitrates for?

EXAM PRACTICE

1. Three gardeners are making their own compost by collecting leaves and putting them in a heap. Bob and Carol both cover their heaps with thick black plastic to absorb heat from the Sun. Bob mixes his by turning it over with a fork every week. Carol leaves hers untouched for six months. Don does not cover his heap but mixes it occasionally.

   a) Who is most likely to produce useable compost first? Give a reason for your answer. **(2 marks)**

   b) Suggest two environmental reasons why it is good to recycle organic waste through composting. **(2 marks)**

   c) Give one economic reason why someone might want to make their own compost. **(1 mark)**

Plant Hormones

Plant hormones are chemicals that control a plant's response to its surroundings, giving it a better chance of survival.

Plants' responses are called **tropisms** and are controlled by **hormones**. They control growth of shoots and roots, flowering and the ripening of fruits. A plant's response to **light** and **gravity** is under the control of auxin.

Control of Growth

Plant growth takes place mainly in the **root tip** and **shoot tip**. These areas contain hormones called **auxins**. They move through the plant in solution by diffusion. Auxins **speed up growth in stems** and **slow down growth in roots**.

Gibberellins are similar to auxins – they promote cell division and cell elongation. They can be used to help increase the rate of **seed germination**.

Response to Light

A plant's response to light is called **phototropism**. Plant shoots grow towards the light. Auxin is spread evenly and the shoot grows upwards.

If light comes from one side, auxin accumulates on the shaded side. The auxin makes these cells **elongate** and grow faster, causing **unequal growth**. The result is that the shoot **bends towards** the light.

The shoot is **positively phototropic**.

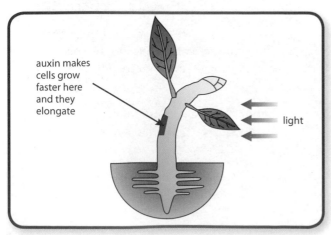

auxin makes cells grow faster here and they elongate

light

Response to Gravity

A plant's response to gravity is called **geotropism**. Even if you plant a seed the wrong way up, the shoot always grows up away from gravity and the root grows down towards gravity.

If a plant is put on its side, auxin gathers on the lower half of the shoot and root. Auxin slows down the growth of root cells so the root curves downwards.

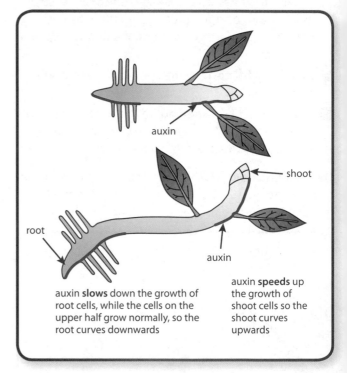

auxin

shoot

root

auxin

auxin **slows** down the growth of root cells, while the cells on the upper half grow normally, so the root curves downwards

auxin **speeds** up the growth of shoot cells so the shoot curves upwards

Auxin causes cells to elongate in the shoot cells, so the shoot curves upwards. The roots are **positively geotropic** and the shoots are **negatively geotropic**.

Rhythms

Organisms respond to regular changes in the environment, for example, woodlands respond to seasonal changes in light intensity.

Photoperiod refers to periods of light and dark, e.g. day length. Photoperiodicity is important in plant germination, growth and reproduction.

Many organisms live with a biological clock (e.g. sleep) that scientists call **circadian rhythms**.

Commercial Uses of Plant Hormones

Growing Cuttings

Rooting powder contains synthetic auxins. A cutting is taken from a plant and dipped in the powder. This stimulates the roots to grow quickly and enables gardeners to grow lots of exact copies (clones) of a particular plant.

Killing Weeds

Synthetic auxins are also used as selective weedkillers, e.g. on lawns. They only affect broad-leaved weeds – narrow-leaved grasses and cereals are not affected. Synthetic auxins kill the weeds by making them grow too fast.

Seedless Fruits

Synthetic auxins are sprayed on unpollinated flowers so the fruits form without fertilisation and without pips, e.g. seedless grapes.

Early Ripening

Plant hormones can also be used to ripen fruit in transport. Bananas and other fruits are picked when they are unripe and less easily damaged. By the time they arrive for sale, they are ripe and ready to eat.

Dormancy

Dormancy keeps seeds from germinating until the conditions are ideal for growth. Hormones can be used to remove the dormancy of the seed to bring about germination at different times of the year.

SUMMARY

- **Hormones in plants control growth, flowering and the ripening of fruits.**
- **Auxin and gibberellins are plant hormones.**
- **Tropisms are plant responses and include phototropism and geotropism.**
- **Hormones can be used commercially: as weedkillers, to produce seedless fruit, to cause early ripening and when growing cuttings.**

QUESTIONS

QUICK TEST

1. What is a plant's response to light called?

2. Does auxin speed up or slow down growth in roots?

3. Does auxin speed up or slow down growth in shoots?

EXAM PRACTICE

1. Bananas are often imported from other countries. They are picked when unripe but ripened before sale.

   a) Why are the bananas picked and transported when unripe? **(1 mark)**

   b) How are they ripened before being sold? **(1 mark)**

2. The diagram illustrates the response of a shoot and a root when a plant is placed on its side. Explain what will happen to both the root and the stem.

   The quality of your written communication will be assessed in your answer to this question.

   (6 marks)

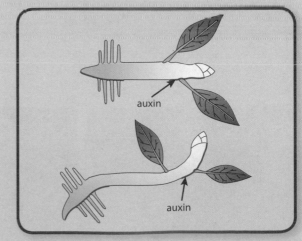
auxin

auxin

Plant Growth

Plants have a general basic structure and grow in a similar way to animals.

Plant Structure

A plant's basic structure is divided up into **five** parts:

1. The **flower** contains the male and female sex organs to make seeds.
2. The **stem** holds the plant upright but water also plays a part in support. The stem contains hollow tubes called **xylem** that carry water and dissolved minerals from the roots to the shoots and leaves. Transpiration is evaporation of water from the leaves. Transpiration plays a part in moving water up the stem. **Phloem** tubes are columns of living cells that carry glucose and other food substances made by the plant up and down the plant to growing and storage areas (called **translocation**).
3. The **root** anchors the plant in the soil and takes up water and minerals.
4. The **root hairs** absorb water and minerals from the soil by increasing the surface area of the roots for more efficient absorption.
5. The **leaf** carries out **photosynthesis** and contains **stomata** (tiny holes) allowing the exchange of gases and water to escape the plant. **Guard cells** open and close the stomata to regulate transpiration.

The loss of water vapour from the leaves drives transpiration. **Transpiration rate** is faster in warm, windy conditions when evaporation of water from the stomata is increased. If plants lose water faster than it is replaced by the roots, the stomata close to prevent wilting.

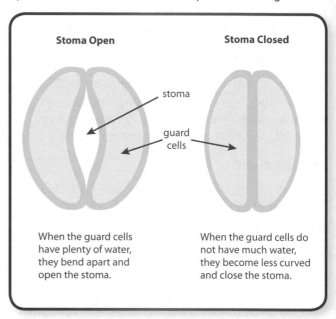

Stoma Open — **Stoma Closed**

stoma

guard cells

When the guard cells have plenty of water, they bend apart and open the stoma.

When the guard cells do not have much water, they become less curved and close the stoma.

Growth

Plants grow by cell division, **mitosis** and **cell differentiation**, but they also undergo a process called **elongation**. This is where the plant cell's vacuole absorbs water and swells. Plant cells also continue to grow, whereas animal cells eventually stop growing.

A plant's root tip and shoot tips are the only areas of a plant that are capable of cell division, so the rest of the plant grows by elongation and cell differentiation. These mitotically active regions are called **meristems**. New cells produced from plant meristems are **unspecialised** and can develop into any kind of plant cell.

Most types of animal cells differentiate at an early stage, whereas many plant cells retain the ability to differentiate throughout life. This is demonstrated when cuttings (clones) are taken from plants. These cuttings are capable, under the right conditions, of growing into an identical plant.

Healthy Growth

Minerals dissolved in water are absorbed from the soil by the roots. Minerals are usually present in the soil in quite low concentrations so **active transport** is used to take them up into root hair cells. This requires energy and can move substances from a low concentration to a high concentration (against a concentration gradient).

If soils are deficient in minerals, then **fertilisers** can be added, usually containing nitrogen, phosphorous and potassium (NPK).

Mineral	Why needed	Deficiency symptoms
Nitrates	To form proteins	Stunted growth, yellow older leaves
Phosphates	Role in photosynthesis, respiration, and making DNA	Poor root growth, purple younger leaves
Potassium	To make enzymes used in respiration and photosynthesis	Yellow leaves with dead spots, poor flower and fruit growth
Magnesium	To make chlorophyll	Yellow leaves

QUESTIONS

QUICK TEST

1. Why do plants need potassium?

2. What would happen to a plant if it could not obtain phosphates from the soil?

3. What is the function of xylem in plants?

EXAM PRACTICE

1. Describe two differences between plant cell growth and animal cell growth. **(2 marks)**

2. When cut flowers are transported, air enters the xylem vessels when the flowers are out of water. This breaks the transpiration stream. Explain why cutting the bottom off the stalks may prolong the life of the flowers once they are placed in a vase of water.

   The quality of your written communication will be assessed in your answer to this question.

   (6 marks)

Animal Growth

Growth is an increase in size or mass of an organism.

The processes of growth in animals include division by **mitosis** and **differentiation**, where cells become specialised for their purpose.

Growth can be measured as an increase in height, wet mass or dry mass.

Multicellular Animals

Multicellular animals have many cells. Being multicellular is advantageous because it allows for…

- an organism to be larger
- cell differentiation
- an organism to be more complex.

Becoming multicellular requires the development of specialised organ systems to allow communication between cells, supplying cells with nutrients and controlling exchanges with the environment.

Variation

Each species of animal has a size range for that particular species. Human height range shows **continuous variation** – it is influenced by a number of genes, not just tall or short. There are many different heights in between the extremes.

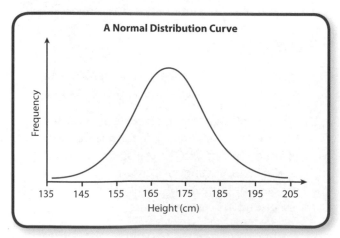

A Normal Distribution Curve

Frequency vs Height (cm): 135, 145, 155, 165, 175, 185, 195, 205

Variation can also be **discontinuous** when values are in discrete groups or categories, for example, blood group (A, B, AB or O), eye colour (blue, green or brown) or gender (male or female).

Variation is caused by our genes, the environment, or a combination of both.

Genetic variation is a result of mutation or reproduction, such as eye colour, blood group or ability to roll your tongue.

Acquired characteristics, such as a tattoo or scar, are caused by the **environment** we live in.

Variations such as height, weight and intelligence are caused by both genetics and environment.

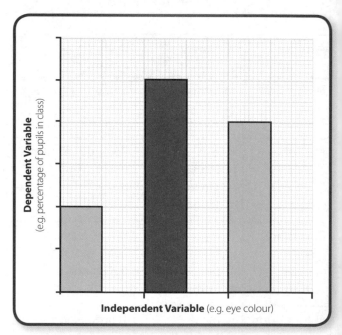

Dependent Variable (e.g. percentage of pupils in class)

Independent Variable (e.g. eye colour)

Human Growth

The main phases of human growth are:

1. Infancy (up to 2 years)

2. Childhood (from 2–11 years)

3. Puberty or adolescence (from 11–13/15 years)

4. Maturity or adulthood (the longest stage)

5. Old age (above 60/65 years)

Humans grow at different rates throughout these stages. In early infancy, the growth is concentrated on the trunk and head, with the arms and legs getting longer by the age of six. This is because during infancy, babies are less mobile and need to concentrate on digestion (in the trunk area) for maximum growth.

During puberty, there is a growth spurt. Then, once maturity is reached, humans actually lose a little height.

Developing foetuses have different growth rates for different parts of their body. At first the head is much larger than the rest of the body, but about half way through pregnancy, the body begins to catch up.

Once born, their weight and head size, if plotted on growth curves, can be an indication of growth problems if outside normal ranges.

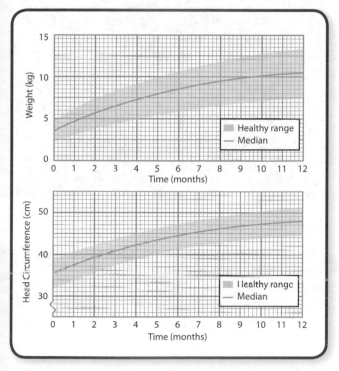

A person's final height and mass is determined by a number of factors including genetics, diet and exercise, hormones, and health and disease.

SUMMARY

● Animal growth includes mitosis and cell differentiation.

● Variation can be continuous or discontinuous and caused by the environment, genetic factors or a combination of both.

● Human growth is split into five phases.

QUESTIONS

QUICK TEST

1. What type of variation is blood group?

2. How do animals grow from a fertilised egg cell?

3. List three factors that influence human height.

EXAM PRACTICE

1. Explain, in detail, why identical twins may be different to each other. **(3 marks)**

2. Humans grow in size and mass because their cells repeatedly divide by mitosis and then specialise.

   a) What are the main phases of human growth? **(5 marks)**

   b) During which phase is there a growth spurt? **(1 mark)**

3. State if the following features are caused by the environment, inheritance or both. **(4 marks)**

   a) Blood group

   b) Long hair

   c) Being good at sport

   d) Colour blindness

Behaviour

Animals inherit patterns of behaviour from their parents, known as instinctive or **innate** behaviour. This is important for **survival** of the species.

Animals can also learn behaviours that protect them from danger, such as knowing when to flee from predators if they hear a certain noise.

Habituation can be a learning process when the animal learns to no longer respond to a stimulus if there is no real danger. Tap the ground by a snail and it will retreat into its shell, but repeat this 10 times and the snail will learn that there is no real danger and ignore the tapping.

Animals can learn through **classical conditioning**, such as Pavlov's observations of dogs producing saliva with the anticipation of food signalled by a secondary stimulus (e.g. ringing of a bell).

Operant conditioning occurs when behaviour is learned due to a pleasurable outcome. An example is using a **Skinner box** – rats press a particular lever and they receive food. Humans can make use of conditioning when training captive animals for specific purposes, including sniffer dogs, police horses and dolphins.

Imprinting is when you learn something because of an experience early on in life. For example, Lorenz observed imprinting in geese – a gosling will follow its mother, which is usually the first largest moving object it sees. However, if a similar-sized object is placed in view of the gosling before it sees its mother, it will imprint on that instead.

An **ethologist** is a person who studies behaviour. Niko Tinbergen is famous for studying innate behaviour in gull chicks where he investigated their begging behaviour – an instinct to peck at a red spot on the adults' lower bill.

In order for Dian **Fossey** to observe social behaviour in gorillas she needed to recognise individual gorillas. The gorillas had to become habituated to her presence so she could get close enough to study them properly. To achieve this she imitated gorilla behaviours and their sounds.

Jane **Goodall** is famous for observing behaviours in chimpanzees and was the first to witness them eating meat, disproving the belief that chimpanzees were vegetarian.

Communication

Communication between animals can happen in many different ways, such as sounds, signals (gestures, body language and facial expressions) and chemicals (pheromones).

Humans can communicate simple ideas with displays of emotion. Humans have also developed more complex levels of communication.

Feeding Behaviours

Plant material is low in nutrients, particularly amino acids, so **herbivores** have to feed for longer. They also tend to feed in herds for protection. Feeding in herds also means there are more ears listening for predators. Herd animals move around to avoid predators and to find new areas of food.

Carnivores feed on protein-rich food and spend less time eating. Predator carnivores have to be good at detecting and catching their food. Some hunt efficiently in packs, like lions, working together to bring down food, while some hunt alone, e.g. bears.

Reproductive Behaviours

Sexual reproduction involves finding and selecting a mate. Some species develop courting behaviour, such as the peacock's display of its tail.

The biggest and strongest animals in a pack will tend to be successful in securing a mate; this ensures that their genes are passed on. Some animals will mate for life, e.g. penguins, whilst other animals will select several mates over a lifetime. Some animals will select a mate for a breeding season, whilst others select different mates throughout a breeding season, e.g. chimpanzees.

Birds and mammals have developed special behaviours to rear their young. This is known as **parental care** – they feed them and care for them until they are able to fend for themselves. This ensures that the parents' genes survive and is important for evolution. Within the animal kingdom, parental care may be risky to the parents.

SUMMARY

- Animals inherit behaviour, known as innate behaviour, in order to help them survive.
- Other behaviours that can be learned include habituation, classical and operant conditioning, and imprinting.
- Animals communicate using sounds, signals and chemicals.
- Animals also show feeding and reproductive behaviours.

QUESTIONS

QUICK TEST

1. What does an ethologist do?

2. What is imprinting?

3. Describe the differences in feeding behaviours between herbivores and carnivores.

EXAM PRACTICE

1. Match the types of learning to the correct example.   (4 marks)

Innate	Horses not frightened by loud sounds
Habituation	Recognising parents
Imprinting	Dogs salivating when bell rings
Conditioning	Foraging for food

2. What is trial and reward learning also known as? Circle the correct answer.

   Innate behaviour   **Learned behaviour**   **Operant conditioning**   (1 mark)

3. Humans show parental care when looking after their young. Discuss what parental care involves, why it is important and the risks involved.   (3 marks)

Food Production

Food Production

Farming has become more intensive to try and provide more food from a given area of land. This reduces the need to destroy the countryside for extra space for farming.

At each stage in a food chain, less material and less energy are contained in the biomass of the organisms. This means that the **efficiency** of food production can be improved by reducing the number of stages in food chains.

Food production can also be more efficient by reducing energy loss from animals. For example, **restricting movement** and **controlling temperature** of their surroundings means energy won't be 'wasted' in movement and keeping the animals warm. The energy will go towards growth and therefore faster weight gain. **Antibiotics** can be used in feed to keep disease at bay.

Intensive Farming

Intensive farming methods mean less labour costs, which equals cheaper produce. Intensive farming may be efficient, but it raises ethical dilemmas. Many people regard intensive farming as cruel to animals as they are confined in small spaces. There is worry about the overuse of antibiotics because of **resistance** and fear of them entering the food chain.

Examples of intensive farming include **battery hens** and **fish farming**.

Fish farming is a way to respond to dangerously low fish stocks due to over fishing. It is important to maintain fish stocks at a level where breeding continues, or certain species may disappear altogether in some areas. Net size and fishing quotas play an important role in conservation of fish stocks.

Organic Farming

Organic farming produces less food per area of land and can be expensive as it is labour intensive. Organic farming uses compost and manure as a **natural fertiliser**. Crops are rotated to improve soil fertility. Weeding is done by hand and pests are managed by **biological control**.

Battery Hen Farming

Fish Farming

Intensive Farming of Plants

Hydroponics is the growth of plants without soil in a special medium. This is useful in areas where soil is infertile. The plants need support and carefully controlled mineral levels. They are grown in greenhouses where factors affecting growth and diseases can be controlled. Tomato plants are often grown in this way.

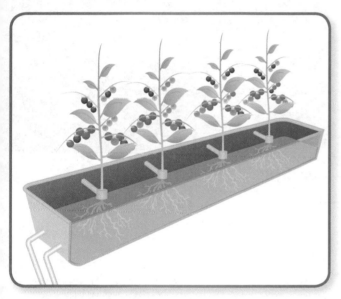

Controlling Pests

Pesticides are used to kill insects (insecticides), fungi (fungicide) and weeds (herbicides). Pesticides can help improve **crop yield**, but there are disadvantages of using pesticides that affect the environment and our health.

Pesticides may enter and accumulate in food chains (e.g. **DDT**), they may harm organisms that are not pests and they can persist in the environment for a long time.

Biological control is an alternative to pesticides. It involves releasing a natural predator of the pest. For example, ladybirds are often used to biologically control greenfly (aphids).

Disadvantages of biological control include that the predator may not eat the pest, it may eat useful species, it may increase out of control, and it may not stay in the area where it is needed.

SUMMARY

- Intensive farming is used to produce cheap food at high yields.
- Pests can be controlled using pesticides or biological control.

QUESTIONS

QUICK TEST

1. Why do farmers that farm intensively restrict the movement of animals?

2. Give one example of intensive farming in animals.

3. Give one example of intensive farming in plants.

EXAM PRACTICE

1. Give two advantages and disadvantages of organic farming and intensive farming.   **(4 marks)**

2. Give one advantage and one disadvantage of using biological control methods.   **(2 marks)**

Microorganisms and Fuels

Most developing countries are short of fuels. An alternative fuel that is cheap and readily available can be produced from **the waste** from processing vegetables, animals and even humans.

Biogas

Biological waste can be dried and burnt, or fermented by bacteria to produce a **biogas** for cooking, heating and lighting and, on a large scale, generating electricity.

Biogas can be produced on a much larger scale using waste from factories or sewage works. Many different microorganisms are involved in the breakdown of materials to produce biogas.

Biogas is seen as a 'cleaner' fuel than diesel and petrol, but it does not contain as much energy as natural gas.

A Biogas Generator

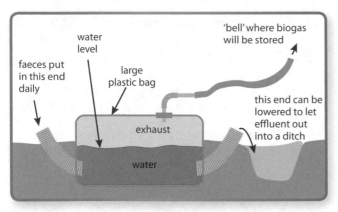

1. Waste is put into the biogas generator.
2. The generator is tightly covered so no air can get in.
3. Inside, bacteria respire **anaerobically** using the waste material as a source of carbohydrate and producing a mixture of carbon dioxide and mainly **methane** (biogas)(more than 50%). There may be traces of hydrogen, nitrogen and hydrogen sulfide.
4. The biogas rises and collects at the top of the generator and can flow along pipes to wherever it is needed; this could be enough to supply a small farm.
5. The waste material left at the bottom of the generator is rich in nutrients and can be used as a fertiliser.

The output of a biogas generator can be affected by climatic conditions.

A Biogas Generator

Landfill

Methane gas is released from landfill sites. It can be dangerous as it is flammable and can explode.

Bioethanol

Fossil fuels can be replaced by ethanol-based fuels produced by the **anaerobic fermentation** of sugar cane and glucose from maize starch.

Brazil was one of the first countries to produce ethanol for fuel using their vast amounts of sugar cane. The juices and waste from the sugar cane can be fermented by yeast in anaerobic respiration to produce **ethanol**.

A mixture of petrol and bioethanol is called gasohol. It is used by cars in some countries such as Brazil. Gasohol is more economically viable in countries that have ample sugar supplies and small oil reserves. Cars in the UK could be modified to use gasohol in the future.

Bioethanol or biofuels are **renewable** and during their production they use carbon dioxide from the atmosphere. The disadvantages of biofuels are that growing crops to make them uses land, which may affect the availability of land to grow food, or even lead to habitat loss and extinction of species.

Global Warming

Levels of carbon dioxide and methane in the atmosphere are increasing and contribute to 'global warming'. An increase in the Earth's temperature of only a few degrees Celsius may cause:

- changes in the Earth's climate
- a rise in sea level
- a reduction in biodiversity
- changes in migration patterns, e.g. in birds
- changes in the distribution of species.

Carbon dioxide can be **sequestered** in oceans, lakes and ponds. This means it is taken from the atmosphere and stored in the water. This is an important factor in removing carbon dioxide from the atmosphere.

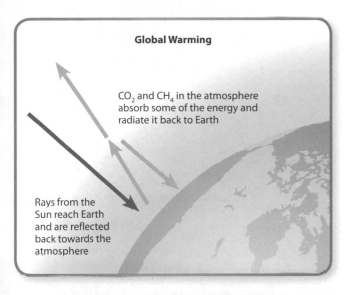

Global Warming

CO_2 and CH_4 in the atmosphere absorb some of the energy and radiate it back to Earth

Rays from the Sun reach Earth and are reflected back towards the atmosphere

SUMMARY

- **Waste products from animals, humans and processing vegetables can be fermented and used as biofuel.**

- **Biogas generators can be used on small and large scales.**

- **Bioethanol is produced by anaerobic fermentation.**

- **Atmospheric levels of CO_2 and methane are increasing, contributing to global warming.**

QUESTIONS

QUICK TEST

1. What is bioethanol?

2. Which type of microorganism is used in a biogas generator?

3. Suggest a use for the waste material left at the bottom of a biogas generator.

EXAM PRACTICE

1. A school decided to set up a biogas generator producing fuel using organic matter and microorganisms.

   a) i) Name the gases produced.   **(1 mark)**

      ii) What could the school use the biogas for?   **(2 marks)**

      iii) What type of respiration is carried out in the fermentation process?   **(1 mark)**

   b) The partially completed table below shows the income and the costs for the biogas generator.

	Yearly cost (£)	Yearly income (£)
Electricity generated from the biogas	_____	26 000
Heating from burning the biogas	_____	3000
Sale of waste after biogas production	_____	7000
Operation and maintenance costs	11 000	_____

   i) Calculate the yearly profit for this biogas generator. Show your calculations.   **(2 marks)**

   ii) The biogas generator cost £250 000 to build and set up. Calculate the number of years it will take to pay back the cost of the generator.   **(1 mark)**

Yeast

Yeast is a single-celled fungus. It reproduces asexually by budding out of the cell.

Yeast feeds on sugars and can respire anaerobically without oxygen producing carbon dioxide and alcohol (ethanol). This is called **fermentation**. Yeast can also respire with oxygen (aerobic).

The equation for anaerobic respiration in plant cells and some microorganisms, including yeast, is shown below:

glucose $\rightarrow$ ethanol + carbon dioxide (+ energy)

$$C_6H_{12}O_6 \rightarrow 2C_2H_5OH + 2CO_2$$

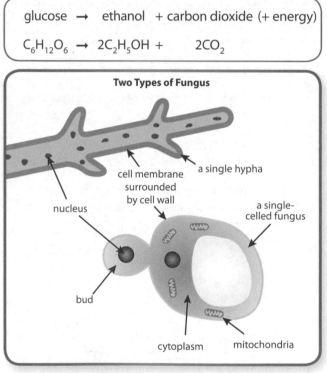

Two Types of Fungus

a single hypha

cell membrane surrounded by cell wall

nucleus

a single-celled fungus

bud

cytoplasm

mitochondria

Food Production

Yeast is used in food substances such as bread. The carbon dioxide that the yeast produces makes the bread rise. The antibiotic **penicillin** (from the fungus *Penicillium*) is produced by **fermentation**.

Microorganisms can be grown (or cultivated) in large quantities to produce the desired product (biomolecules) in a large metal vessel called a **fermenter**. Products include antibiotics and other medicines, proteins, enzymes for food processing and commercial products.

Biotechnology is described as the alteration of natural biomolecules using science and engineering to provide goods and services.

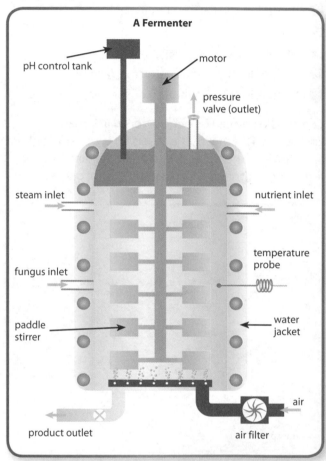

A Fermenter

pH control tank

motor

pressure valve (outlet)

steam inlet

nutrient inlet

temperature probe

fungus inlet

paddle stirrer

water jacket

air

product outlet

air filter

To achieve the desired product as cheaply and efficiently as possible, the following procedures need to be followed:

● **Temperature** needs to be maintained – microorganisms have an optimum temperature at which they and their enzymes work best. Yeast growth rate doubles for every 10°C rise up to a maximum temperature (when it is too hot the yeast will be killed). The water-cooled jacket removes excess heat from the process.

● **pH**, **oxygen** and **nutrient** levels need to be maintained – sensors in the fermentation vessel achieve this and feedback where necessary.

- The vessel must be kept **sterile**; the entry of unwanted substances would spoil the end product. The air going into the vessel is filtered to remove unwanted bacteria and other unwanted substances.
- The microorganisms need to be kept supplied with **sufficient substrate** (food). The yeast needs sugar for energy and proteins for growth.
- The vessel has a **stirrer** to ensure that the substrate and microorganisms are kept in close contact.

Here are examples of food substances made in a fermenter:

- **Mycoprotein** is a protein-rich food (high in fibre and low in fat). It is made by growing the *Fusarium* fungus (it feeds on starch). The mould grows using up all the nutrients in the fermenter before it starts making mycoprotein.
- **Beer** is made from the grains of barley. When the barley grains germinate, the starch in the grain is broken down into a sugary solution by enzymes (this is called malting). The malted barley is mashed, heated with water to dissolve the sugar, cooled and then added to yeast in the fermenter. The yeast breaks down the sugar into alcohol. Hops are added to give the beer flavour. The beer is then put into casks, kegs or bottles after it has been pasteurised.
- **Wine** is made in a similar way to beer but grapes are used instead of barley. The grapes are crushed releasing their sugar.

Using microorganisms for food production has its benefits, including:

- rapid population growth
- being easy to manipulate
- production is independent of climate
- waste products can be used from other industrial processes.

Fermentation is limited because the increasing amounts of alcohol produced stop some yeasts from growing – this puts a limit on the alcohol content of beer and wine.

However, different strains of yeast can tolerate different levels of alcohol.

To make spirits, the wine or beer can be distilled. This is a commercial process and needs licensed premises. Whisky can be made by distilling malted barley.

SUMMARY

- **Yeast is a single-celled fungus, which reproduces asexually and can respire anaerobically.**
- **Fermentation is a type of anaerobic respiration that produces carbon dioxide and alcohol.**
- **Yeast is involved in food production using fermenters.**

QUESTIONS

QUICK TEST

1. What is a fermenter?

2. Why are hops added to beer?

3. What is mycoprotein?

EXAM PRACTICE

1. Yeast is used to make beer in a process called fermentation.

   a) What is fermentation? **(1 mark)**

   b) Describe the main stages in producing beer. **(5 marks)**

   c) Give two other food products that can be made by yeast in a fermenter. **(2 marks)**

Bacteria

Bacterial cells are smaller and simpler than plant and animal cells. They are typically only a few thousandths of a millimetre (mm) in size.

Bacteria are thought to be the earliest forms of life. Bacteria reproduce rapidly by a type of asexual reproduction called **binary fission**. Their rapid growth can spread disease quickly and contaminate food substances, but they are also useful in the production of food and composting. There are three main shapes of bacteria, as shown in the diagram below:

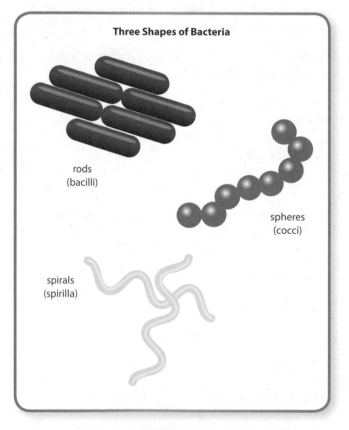

Three Shapes of Bacteria

rods
(bacilli)

spheres
(cocci)

spirals
(spirilla)

Bacteria are ideal for industrial and genetic uses for various reasons:

- They can reproduce quickly.
- They have plasmids that are easy to genetically manipulate.
- Their biochemistry is simple.
- They are able to make complex molecules.
- There are few ethical concerns over their use.

Food Production

- **Vitamin C** – produced by *Acetobacter* species of bacteria.
- **Amino acids and monosodium glutamate (MSG)** – produced by *Corynebacterium glutamicum* bacterium.
- **Cheese** – produced by the action of *Lactobacillus lactis* bacteria on milk. First the milk is warmed and a starter culture of bacteria is added. Modern-day cheese is made using an enzyme manufactured by genetically engineered microorganisms called chymosin. Chymosin is added which speeds up the milk separation process into curds (lumps of protein and fat) and whey (a liquid). The curds are pressed together to make cheese. Different bacteria cultures and some moulds are used to make different cheeses.
- **Yoghurt** – also produced by the action of bacteria on milk. The milk is heated (pasteurised) to remove unwanted microorganisms and then cooled so that a starter culture containing two types of bacteria can be added. The bacteria **ferment** the sugar called **lactose** in the milk and change it into **lactic acid**. The lactic acid gives yoghurt a sharp taste and makes the milk clot, producing thick yoghurt. Fruit can be added to the yoghurt for flavour.

Aseptic Techniques

Cultures of microorganisms are needed for investigating the action of disinfectants and antibiotics. Microorganisms such as bacteria and fungi can be grown on a special jelly called **agar**. Agar is called a **nutrient medium** and can be poured into a Petri dish.

When growing cultures it is important that they are uncontaminated and that a safe technique is followed (**aseptic techniques**):

- The Petri dish used must be sterilised before use to kill unwanted microorganisms.
- The agar is boiled, poured into a Petri dish and allowed to cool; the lid is only lifted briefly to add the microorganisms.
- An inoculating loop is used to transfer the microorganisms to the agar and this must also be sterilised by passing it through a flame.
- The Petri dish lid must be sealed with tape to prevent contamination from microorganisms in the air.
- Cultures in schools and colleges are then incubated at a maximum temperature of 25°C – this reduces the likelihood of harmful microorganisms growing. In industry, higher temperatures can be used to produce more rapid growth.

Louis **Pasteur** contributed to the development of aseptic techniques.

Preserving Food

Key factors in the process of decay include the presence of microorganisms, a warm temperature, oxygen and moisture. Food manufacturers use various techniques to preserve food and to reduce the rate of decay by eliminating these key factors.

Techniques include canning, cooling, freezing, drying, adding salt or sugar and adding vinegar.

Nanotechnology involves manipulating structures that are about the same size as some molecules. Nanotechnology is used in the food industry, including food packaging that can increase shelf life and detect contaminants by changing colour when the products of food spoilage are detected.

SUMMARY

- **Bacteria reproduce by asexual reproduction.**
- **Some bacteria are used in food production.**
- **Aseptic techniques help to grow microorganisms in sterile conditions.**
- **Various techniques can be used to preserve food.**

QUESTIONS

QUICK TEST

1. Why are schools not allowed to grow microorganisms above 25°C?

2. Name the enzyme produced by genetically engineered organisms in cheese production.

3. Name the three main shapes of bacteria.

EXAM PRACTICE

1. The starter culture added to milk to make yoghurt contains *Lactobacillus bulgaricus* and *Streptococcus thermophilus*.

   a) What type of microorganisms are these?
   (1 mark)

   b) Why is milk heated first when making yoghurt?
   (1 mark)

   c) Why is it cooled before adding the starter culture?
   (1 mark)

   d) What is the product of fermentation and what is its effect on milk?
   (2 marks)

2. Which of the following uses knowledge of osmosis to preserve food? Circle the correct answer.
   (1 mark)

   Canning   Salting   Cooking   Freezing

Genetic Modification

Genetic modification is where a gene from one organism is transferred to another and continues to work.

There have been great developments in genetic modification technology, but there is still a lot that remains unknown. The completion of the human genome project and the mapping of genes from plants and animals have increased the possibilities of research.

The benefits of genetically modified (GM) crops are that crop characteristics can be improved quickly, such as resistance to pests. For example, a toxin produced by *Bacillus thuringiensis* bacterium is inserted into plants and makes them insect resistant.

Agrobacterium tumefaciens is a bacterium that is commonly used as a vector to transfer gene coding for herbicide resistance into a plant cell such as soya beans. Increasing resistance to herbicides may increase the crop yield due to reduced competition.

The main steps in genetic modification include the following:

1. Isolating and replicating the required gene.

2. Putting the gene into a suitable vector (virus or plasmid).

3. Using the vector to insert the gene into a new cell.

4. Detecting individuals

5. Selecting the modified individuals.

Making Insulin

Many disorders are caused when the body cannot make a particular protein.

Genetic engineering has been used to treat people with **diabetes** by the production of the protein, **insulin**. This involves the use of **plasmids** (circular DNA) from bacteria.

The gene that codes for insulin is cut out by special enzymes (**restriction enzymes**) from human pancreas cells and then inserted into a plasmid. Cutting open DNA leaves 'sticky ends' that allow ligase enzymes to rejoin DNA strands. The modified plasmid is put back into the bacterium and it is allowed to multiply in a fermenter where it produces insulin.

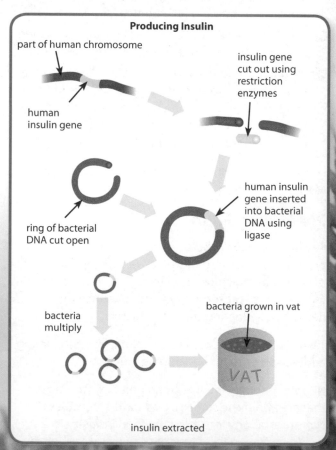

Producing Insulin

part of human chromosome

human insulin gene

insulin gene cut out using restriction enzymes

ring of bacterial DNA cut open

human insulin gene inserted into bacterial DNA using ligase

bacteria multiply

bacteria grown in vat

VAT

insulin extracted

Risks

The risks involved using GM crops include possible allergic reactions and the possibility of the spread of pest and herbicide resistance to wild varieties of crops and weeds.

Some people argue that genetic modification is safe and is necessary in order to match population growth and help feed millions in developing countries more effectively. Others argue that there is more than enough food in the world and the problem is the distribution of food, not its production.

It is important to collect reliable data about the use of GM crops (e.g. the use of farm-scale field trials) in order that possible effects on the environment and on health are understood. The data may then be used to help develop policies about the use of these crops and inform consumers. Information should be unbiased as it could affect public perception of foods containing GM products.

SUMMARY

● **Genetic modification is where a gene from one organism is transferred to another and still works.**

● **Bacteria are genetically modified to produce insulin.**

● **There are risks involved with using GM products.**

QUESTIONS

QUICK TEST

1. Which bacterium is used as a vector to give herbicide resistance to plants?

2. What are circular pieces of DNA found in bacteria called?

3. What is used to cut out genes?

EXAM PRACTICE

1. The following sentences describe how genetic engineering occurs. Put them in the correct order. **(1 mark)**

   A – DNA ligase joins the gene into the plasmid

   B – The bacteria are grown in a fermenter

   C – Restriction enzymes cut the DNA and plasmid

   D – The gene is identified

   E – The bacteria are tested to ensure they have taken up the plasmid containing the desired gene

   F – The plasmid is removed from a bacterium cell

2. Where is the human gene for insulin expressed? **(1 mark)**

3. Name the potential problem with using bacteria for genetic engineering. **(1 mark)**

Answers

Pages 4–5
QUICK TEST
1. The variable we choose to change in an experiment; 2. The variable that we measure in an experiment; 3. An irregular value that does not fit the general pattern or trend; 4. A variable that can be listed in order, e.g. small, medium, large; 5. A bar graph
EXAM PRACTICE
1. a) The time of day (1)
   b) The water lost (mm³) by the plant (1)
   c) Anomalous results are 10:00 – 105 mm³ (1); Does not fit the trend, less than expected (1)
   d) 12:00 (1)
   e) 6 + 5 + 7 + 55 + 125 + 105 + 248 + 230 + 128 + 112 + 8 + 7 = 1036 mm³ (1)
   f) Repeat the investigation another day or with other similar plants (1)
   g) Line graph with a curve of best fit (1)

Pages 6–7
QUICK TEST
1. Carbohydrates, proteins, fats, vitamins, minerals, fibre and water;
2. $BMI = \dfrac{mass\ (kg)}{(height\ (m))^2}$; 3. Statins
EXAM PRACTICE
1. Saturated fats (1)
2. The rate at which chemical reactions take place in the body (1)
3. Needed for: Making cell membranes (1), Insulation (1), Providing energy (1).

Pages 8–9
QUICK TEST
1. A group of cells with similar structure and function; 2. Muscular, glandular and epithelial tissue; 3. Protease, lipase and carbohydrase enzymes
EXAM PRACTICE
1. mouth → oesophagus → stomach → small intestine → large intestine → anus (1)
2. Emulsifies fats (1); Provides alkaline conditions for enzymes (1)
3. Shortage of donors (1); Correct tissue match (1); Size and age (1)
4. The drugs prevent the organ being rejected by the body (1); The immune system will not produce antibodies against the new 'foreign' tissue (1)

Pages 10–11
QUICK TEST
1. Amino acids; 2. Lipase; 3. **Accept suitable answer, e.g.:** Pre-digest protein in baby food
EXAM PRACTICE
1. **Key ideas:** Biological catalysts; Speed up reactions; Lock and key mechanism; Reduce temperature and pressure of reactions; Save energy and money
   Model answer: Enzymes are biological catalysts that speed up the rate of reactions. They work by a lock and key mechanism where a substrate fits into an active site. Enzymes are used in industry because they help reactions to happen at lower temperatures and pressures. This helps to save energy and money for the manufacturer.
2. a) For a fair test (1); To enable a comparison to be made (1)
   b) **Any one from:** Catalase / enzyme in liver more effective; Liver contains more enzymes; Conditions favoured the liver enzymes (1)

Pages 12–13
QUICK TEST
1. glucose + oxygen → carbon dioxide + water (+ energy); 2. Lactic acid and energy; 3. The amount of oxygen needed to oxidise the lactic acid to carbon dioxide and water
EXAM PRACTICE
1. a) A (1)
   b) **Any two from:** The recovery rate was quicker/got back to resting rate sooner; Lower starting pulse rate; Lower pulse rate during exercise; Lower pulse rate immediately after exercise. (2)
   c) Heart rate monitors (1); **Any two from:** Takes a continuous measurement; Less chance of human error. (1)

Pages 14–15
QUICK TEST
1. Cartilage acts as a shock absorber; Synovial fluid reduces friction;
2. Long bones are hollow; 3. Shark
EXAM PRACTICE
1. a) Ball and socket (1)
   b) One muscle contracts (e.g. biceps) and the other relaxes (e.g. triceps) (1)
   c) Antagonistic pairs (1)

Pages 16–17
QUICK TEST
1. Motor, sensory and relay; 2. Junction between two neurones; 3. Light
EXAM PRACTICE
1. receptor (1); motor (1); effector (1)
2. By transmission/diffusion of a chemical (neurotransmitter) (1); Across a synapse (1)
3. Myelin is a fatty sheath (1); insulating the neurone from neighbouring cells (1); which increases the speed of impulse (1)

Pages 18–19
QUICK TEST
1. Binocular – eyes at the front; Monocular – eyes at the side;
2. Ciliary muscles; 3. Fat and round
EXAM PRACTICE
1. Iris – Controls how much light enters the eye; Lens – Changes shape to focus; Suspensory ligaments – Hold the lens in place (2)
2. Light passes through the cornea and lens (1) and is focused on the retina (1)
3. Stiff suspensory ligaments (1), weak ciliary muscles (1); Causes problems when judging distances (1)

Pages 20–21
QUICK TEST
1. Medulla; 2. Cerebellum; 3. Sensory memory, short-term memory and long-term memory
EXAM PRACTICE
1. Repeating an experience increases the likelihood of nerve pathways transmitting impulses (1)
2. It makes it possible for animals to adapt to new situations (1)

Pages 22–23
QUICK TEST
1. Pancreas; 2. Increases it; 3. Cools the body
EXAM PRACTICE
1. a) **Any two from:** Vasoconstriction of blood vessels; Shivering; No sweat; Increased respiration (2)
   b) Hypothalamus (1)
   c) 37°C (1)
2. **Key ideas:** Pancreas detects blood glucose level too low; Releases glucagon; Travels to liver in blood; Converts glycogen into glucose; Glucose level returns to normal; Example of negative feedback

Model answer: The pancreas detects that the glucose level in the blood is too low. The pancreas releases the hormone glucagon, which travels in the blood to the liver. Glucagon converts stored glycogen in the liver into glucose. The glucose level returns to normal. This is an example of part of a negative feedback system.

Pages 24–25
QUICK TEST
1. Controlling the amount of water in our body; 2. Transplant or dialysis; 3. Pituitary gland in the brain
EXAM PRACTICE
1. a) The liver (1); Excess proteins/amino acids (1)
   b) The blood arrives at the kidney nephron containing urea (1); Urea is forced out of the blood into the Bowman's capsule (1); Not reabsorbed so collects in the collecting duct (1); Flows to the bladder and forms urine (1)
   c) Excess ions (1); Excess water (1)
2. a) Small volume (1); Concentrated (1)
   b) ADH released by the pituitary gland (1); Travels in blood to kidney (1); Causes more reabsorption of water back into the blood (1)

Pages 26–27
QUICK TEST
1. Breathing in and out; 2. The exchange of carbon dioxide and oxygen between the lungs' alveoli and the blood; 3. Two sets of intercostals and the diaphragm
EXAM PRACTICE
1. a) The lungs are a 'dead end' (1)
   b) **Any three from:** Wheezing; Shortness of breath; Chest tightening; Coughing (3)
   c) Mucus traps dust and some microorganisms (1); Ciliated cells line the trachea and bronchi (1); Cells move mucus up and away from the lungs (1)
2. **Any three from:** Large surface area; Excellent blood supply; Moist; Thin (3)

Pages 28–29
QUICK TEST
1. Ovaries; 2. FSH and LH; 3. Release of an egg

EXAM PRACTICE
1. FSH **(1)**; oestrogen **(1)**; LH **(1)**; progesterone **(1)**
2. a) Oestrogen and progesterone **(1)**
   b) FSH **(1)**
   c) Pituitary gland **(1)**

Pages 30–31
QUICK TEST
1. Bacteria, viruses, fungi and protozoa;
2. Transfers pathogens from one organism to another. An example is mosquitoes;
3. Pathogens
EXAM PRACTICE
1. a) Protozoa **(1)**
   b) Animal vector/Mosquito **(1)**; Bites infected humans **(1)**; Sucks in protozoan from blood **(1)**; Bites non-infected person, injecting protozoan **(1)**
   c) **Any two from:** Insect repellent on body; Insect nets; Covering up; Large scale spraying of mosquito repellent in the environment. **(2)**
2. A Athlete's foot **(1)**
3. C Tuberculosis **(1)**

Pages 32–33
QUICK TEST
1. Antitoxins; 2. Antigens on their cell surface; 3. Sterilising equipment, disinfectants, antiseptics and general good hygiene
EXAM PRACTICE
1. The phagocytes can engulf the bacteria **(1)**
2. a) Antigens are found on the surface of pathogens **(1)**; Antibodies are produced by white blood cells. They have a complementary shape to the pathogen's antigens **(1)**
   b) Lymphocytes produce antibodies **(1)**; Phagocytes engulf pathogens **(1)**
3. Antibodies are specific to particular antigens **(1)**

Pages 34–35
QUICK TEST
1. They mutate; 2. Measles, mumps and rubella; 3. *Penicillium notatum*
EXAM PRACTICE
1. Antibiotics are only effective against bacteria **(1)**, influenza is caused by a virus **(1)**
2. a) Contains a dead or harmless pathogen / Still has antigens **(1)**, Triggers antibodies to be made **(1)**
   b) Colds are caused by a different virus **(1)**; The flu vaccine produces antibodies specific to the flu vaccine **(1)**
   c) Don't suffer the disease **(1)**; Helps prevent the spread of disease **(1)**

Pages 36–37
QUICK TEST
1. By blocking the transmission of nerve impulses; 2. Mental illness; 3. Stimulate muscle growth
EXAM PRACTICE
1. **Key ideas:** Equal volumes of caffeinated and decaffeinated cola; Dropping ruler to measure reaction time; Control variables; Repeating to make reliable; Processing results
   Model answer: Measure out equal volumes of the caffeinated cola

and the decaffeinated cola (the independent variable). Use equal volumes to make the test fair. Drink the decaffeinated cola first and wait a certain amount of time (e.g. 5 minutes). Use a 1 m ruler to measure reaction time. Student A holds it above the hand of student B ('0' should be nearest the hand). Student A releases the ruler, student B catches it as quickly as possible. The distance caught is recorded (where first finger is on the scale) and repeated to make results more reliable. This is the dependent variable. Student then drinks caffeinated cola and waits same amount of time. Repeat the dropping of the ruler. Students carry out calculations or use a conversion graph to convert distance into seconds to analyse reaction time. The shortest distance or lowest time equals the fastest reaction time. Caffeine is a stimulant and should increase reaction time.

Pages 38–39
QUICK TEST
1. Both alleles are the same (e.g. RR, rr); 2. A form of reproduction that involves only one parent; Offspring are genetically identical to the parent and each other (clones); 3. Gametes / Sex cells (e.g. sperm, egg)
EXAM PRACTICE
1. a) Tongue rolling = RR or Rr, non-tongue rolling = rr **(1)**
   b)

	R	r
R	RR	Rr
r	Rr	rr

   (1 mark for correctly set out table, 1 mark for correct combination for offspring)
   c) 25% (1 in 4) **(1)**
2. **Key ideas:** Alleles; Recessive/dominant; Genotype/phenotype; Homozygous dominant, homozygous recessive, heterozygous
   Model answer: The two black cats do not have grey fur so the grey allele must be recessive. Both adult cats have the grey allele, but also have a black allele, which is dominant and masks it. The parents are heterozygous (Bb – one dominant black allele and one recessive grey allele). The kittens with grey hair received two grey alleles from each parent during fertilisation. They are homozygous recessive (bb). The kittens with black hair received at least one dominant black allele from a parent during fertilisation. The black kittens could be heterozygous (Bb) or homozygous dominant (BB).

Pages 40–41
QUICK TEST
1. Homozygous recessive, cc; 2. Likely to survive malaria; 3. Dominant
EXAM PRACTICE
1. a) Both parents are non-sufferers; but must be carriers of cystic fibrosis **(1)**; Shaun has cystic fibrosis so he

must have inherited a recessive allele from each parent **(1)**
   b) Heterozygous **(1)**
   c) **Any two from:** Thick mucus; Breathing difficulties; Chest infections; Digestive problems. **(2)**

Pages 42–43
QUICK TEST
1. Where genes from one organism are removed and inserted into cells of another; 2. A technique to correct defective genes responsible for diseases by replacing abnormal genes with normal genes; 3. Crops that have had their genes modified by having new genes inserted
EXAM PRACTICE
1. a) Electrophoresis **(1)**
   b) Identifying all the genes in human DNA and studying them **(1)**
   c) **Any three from:** Identifying suspects; Clearing people who might have been wrongly accused; Identifying paternity; Matching organ donors with recipients. **(3)**
2. **For:** More food supply to starving populations; Less use of harmful insecticides **(2)**; **Against:** Humans against God/nature; Health problems; May pollinate with wild crops **(2)**
3. a) **Advantage** – identify if the unborn child has the genetic disease **(1)**; **Disadvantage** – not always reliable/ there is risk of miscarriage **(1)**
   b) **Any one from:** Parents might need to decide to have a termination; Some people consider this wrong. **(1)**

Pages 44–45
QUICK TEST
1. Asexual reproduction; 2. They are clones of each other but not their parents; 3. She was the first mammal cloned (in 1996)
EXAM PRACTICE
1. a) Body cell **(1)**
   b) So that Dolly did not have too much genetic information **(1)**
   c) Adult cell cloning **(1)**
2. a) Tissue culture **(1)**; Cuttings **(1)**
   b) Cuttings **(1)**; **Any one from:** It is cheap, quick and easy; Doesn't require sterile conditions, which can be expensive. **(1)**

Pages 46–47
QUICK TEST
1. Arteries, the aorta, pulmonary artery; 2. To withstand high pressure; 3. SAN
EXAM PRACTICE
1. **Key ideas:** Diastole; Atria systole; Ventricular systole; Valves
   Model answer: When the heart muscle is relaxed (diastole) the atria fill with blood from the veins. The atria then contract (systole) pushing blood into the ventricles. The ventricles then contract (systole) forcing blood into the arteries. Valves prevent the blood from flowing in the wrong direction. The whole heart relaxes and fills up with blood again.

2. a) Shortage of donor hearts **(1)**; Donor hearts have to be correct match in size, age and tissue **(1)**
   b) There is the risk of rejection **(1)**

Pages 48–49
QUICK TEST
1. Closed; 2. William Harvey; 3. **Accept any three:** Smoking; Stress; Drug abuse; Obesity; Eating saturated fats; High levels of salt; Excess alcohol; Genetic factors
EXAM PRACTICE
1. a) It ensures that blood gets oxygenated **(1)** and that pressure is maintained **(1)**
   b) To deliver food and oxygen **(1)** and remove carbon dioxide and waste substances quickly and efficiently **(1)**
2. a) This is the systolic pressure, which is pressure when the heart contracts **(1)**
   b) **Any two from:** Lower salt and fat content of diet; Exercise more; Reduce alcohol consumption; Don't smoke. **(2)**

Pages 50–51
QUICK TEST
1. Blood group O; 2. Blood group AB; 3. Blood clotting
EXAM PRACTICE
1. a) and b)

Feature	Benefit in the role of carrying oxygen
Biconcave shape	To maximize surface area for diffusion of oxygen
No nucleus	To maximize room for oxygen to be carried
Contains haemoglobin	To maximize ability to carry oxygen

   (1 mark for each correct line)

Pages 52–53
QUICK TEST
1. Pyramid of numbers; 2. Organism that eats the producer; 3. To provide a large surface area to absorb light for photosynthesis.
EXAM PRACTICE
1. a) A – red, atmospheric level of CO_2 **(1)**; B – purple, rate of photosynthesis > respiration in the plants leading to CO_2 in the water being used up **(1)**; C – yellow, plants respiring increases the amount of CO_2 in water. The foil prevents photosynthesis **(1)**
   b) By repeating his investigation **(1)**
   c) The different samples of pond water **(1)**
2. **Any four from:** Flat and thin so there is a large surface area for absorbing sunlight; Thin so CO_2 can reach the cells easily; Stomata on lower surface for gas exchange by diffusion; Veins to carry substances to and from the leaf; Chloroplasts with chlorophyll concentrated near upper surface to absorb sunlight. **(4)**
3. Light, concentration of CO_2 and temperature **(1)**

Pages 54–55
QUICK TEST
1. Sulfur dioxide and nitrogen oxides; **2.** Carbon dioxide and methane; **3.** Aquatic invertebrate, mayfly nymph
EXAM PRACTICE
1. Key ideas: Fertilisers increase algal growth which blocks light; Reduce photosynthesis; Plants die and bacteria decompose them; Bacteria populations increase, use up oxygen; Aquatic organisms die due to lack of oxygen; Reduced biodiversity
Model answer: Fertilisers used by farmers run off the land into watercourses where they increase algal growth. This blocks the light to plants below the surface of the water, restricting photosynthesis for the plants there. These plants then die and bacteria decompose them. The bacteria populations increase and use up the available oxygen in the water. Aquatic organisms can no longer survive in the water as there is not enough oxygen. These organisms die, which reduces the biodiversity.
2. a) Chopping down trees **(1)**
   b) Accept two suitable answers, e.g.: For timber; For agriculture, For roads; Space for buildings. **(2)**
   c) Accept two suitable answers, e.g.: Soil erosion; Habitat loss; Less rainfall; Less food; Disruption to food chains. **(2)**
3. More trees available to absorb CO_2 from the air **(1)**; Less burning of timber which increases CO_2 in the air **(1)**

Pages 56–57
QUICK TEST
1. Meeting the needs of today without damaging the Earth for future generations; **2.** The number and genetic variations of species and within species; **3.** When organisms interact and rely on each other for life
EXAM PRACTICE
1. a) Bacteria use hydrogen sulfide and oxygen to make food **(1)**; Their proteins are able to withstand high temperatures **(1)**
   b) Antarctic **(1)**; High altitudes **(1)**
   c) Photosynthesis uses light to make glucose **(1)**; Chemosynthesis uses chemicals to make glucose **(1)**

Pages 58–59
QUICK TEST
1. Kingdom, phylum, class, order, family, genus, species **2.** There are too many organisms to count; Saves time; **3.** No deaths, immigration or emigration, identical sampling methods and marking not affecting survival rate.
EXAM PRACTICE
1. a) Quadrat **(1)**
   b) Set an area **(1)**; Place quadrats randomly within the area **(1)**; Count the number of different plant species **(1)**; Repeat several times **(1)**
   c) Transects **(1)**
   d) To help accurately identify organisms found **(1)**

Pages 60–61
QUICK TEST
1. A group of living organisms able to breed together to produce fertile offspring; **2.** 5; **3.** Animals with a backbone – Phylum Chordata
EXAM PRACTICE
1. a) Infertile **(1)**; Lions and tigers are considered separate species, so hybrids are unlikely to be fertile **(1)**
   b) *Panthera* **(1)**
2. habitat **(1)**
3. They are able to make their own food by photosynthesis **(1)**

Pages 62–63
QUICK TEST
1. Where an organism lives that provides it with the conditions necessary for survival; **2.** The predator / prey cycle keeps numbers constant; **3.** Competition between individuals of the same species.
EXAM PRACTICE
1. Key ideas: Arctic fox: Small ears – reduces surface area to conserve heat loss, Thick fur – for insulation; White – camouflage in snow;
Fennec fox: Large ears – increases surface area to increase heat loss; Thinner fur – reduces insulation; Brown/sandy colour – camouflage in desert
Model answer: The arctic fox has small ears with a small surface area so less heat is lost. The fennec fox has large ears with a large surface area so more heat is lost. The arctic fox has thick white fur for insulation and camouflage in the snow. The fennec fox has thinner fur in order to reduce insulation and is a brown / sandy colour so is camouflaged in the desert.
2. a) A man-made habitat and climate **(1)**
   b) Tight control of the physical and chemical components of the ecosystem would mean there is less variety in the habitat **(1)** and so there is less likely to be variety in the biological organisms that inhabit it **(1)**

Pages 64–65
QUICK TEST
1. Only the best adapted to a change in the environment will survive and breed; **2.** Suggested changes that occur in an organism during its lifetime can be inherited; **3.** The preserved remains of dead organisms
EXAM PRACTICE
1. Any two from: Challenged idea of God; Insufficient evidence; Mechanisms of inheritance unknown. **(2)**
2. Key ideas: Variation within species; Competition; Best suited survive; Breed and pass on genes; Less adapted don't survive; Extinct
Model answer: Darwin's theory is that there is great variation within species, and species compete for resources (food, water, mate, etc). The individuals within the species that are the best suited/adapted will be more successful and will survive. They therefore are able to breed and pass

on their advantageous genes to their offspring. This is known as 'survival of the fittest'. Those individuals that are not well adapted are less likely to survive and may become extinct.

Pages 66–67
QUICK TEST
1. Respiration; Release of energy from glucose; **2.** Long and thin, which increases the surface area to absorb water and minerals; **3.** A cell that has the ability to replicate and differentiate into different types of tissue throughout the life of the organism
EXAM PRACTICE
1. Chloroplast **(1)**; Vacuole **(1)**; Cell wall **(1)**
2. Key ideas: For: Stem cells could be used to save/improve lives; Alternative to mechanical devices and donor organs; **Against:** It's not right to interfere with nature; Expensive; Risks to human health
Model answer: There are several arguments for and against stem cell therapy. Some people argue that stem cells can be used to save/improve lives, for example growing organs that can be used as an alternative to mechanical devices and donor organs. However, other people have a moral objection and say that it's not right to interfere with nature. Stem cell therapy is also expensive and poses some risks to human health.

Pages 68–69
QUICK TEST
1. The movement of particles from an area of high concentration to an area of low concentration; **2.** Allows small molecules to pass through, but not large ones; **3.** They have no cell wall
EXAM PRACTICE
1. Thin walled – less diffusion distance **(1)**; Lots of them – large surface area **(1)**; Close contact with blood vessels – short diffusion distance **(1)**; Blood takes oxygen away – sets up a concentration gradient **(1)**
2. a) diffusion **(1)**; **b)** diffusion **(1)**; **c)** diffusion **(1)**; **d)** osmosis **(1)**; **e)** diffusion **(1)**
3. a) 0M **(1)**
   b) Water entered the cells **(1)** through the partially permeable membrane by the process of osmosis **(1)**
   c) Repeat the experiment **(1)**
   d) Use a more sensitive balance **(1)**

Pages 70–71
QUICK TEST
1. Full/normal number of chromosomes, i.e. 46; **2.** Mutation; **3.** Oviduct
EXAM PRACTICE
1. F, D, A, B, E, C **(1)**
2. During fertilisation **(1)**; Sperm (23) + egg (23) = 46 **(1)**

Pages 72–73
QUICK TEST
1. 2; **2.** 46; **3.** Long strands of DNA
EXAM PRACTICE
1. a) meiosis **(1)**; **b)** mitosis **(1)**; **c)** meiosis **(1)**;
2. a) True **(1)**; **b)** False, 46 **(1)**; **c)** False, 4 **(1)**

Pages 74–75
QUICK TEST
1. Photosynthesis; **2.** Respiration and burning / combustion; **3.** Making proteins
EXAM PRACTICE
1. a) Bob **(1)**; **Any one from:** His compost will be warm (black plastic) and will get oxygen because it is turned over; These are good conditions for bacteria to grow and digest the waste. **(1)**
   b) Any two from: Organic waste in landfill produces methane (greenhouse gas); Recycling means less space taken up by landfill sites; Reduces the number of people buying peat-based compost (destruction of peat bogs is an environmental issue). **(2)**
   c) Any one from: The compost can be used on plants avoiding the need to buy compost or fertilisers; Increasing yield of crops. **(1)**

Pages 76–77
QUICK TEST
1. Phototropism; **2.** Slow down; **3.** Speed up
EXAM PRACTICE
1. a) Any one from: Less chance of damage in storage; Last longer; Easier to pick. **(1)**
   b) They are sprayed with plant hormones **(1)**
2. Key ideas: Roots – Auxin on underside of root due to gravity; Inhibits the growth of cells; Upper side grows faster than the underside; Root grows downwards – positive geotropism; **Shoots** – Auxin causes more growth on underside (no sunlight); Shoot bends upwards towards the light – positive phototropism.
Model answer: When a plant is placed on its side, auxin gathers on the underside. In the root, the auxin inhibits the growth of cells, so the upper side grows faster than the underside, causing the root to grow downwards in the direction of gravity. This is called positive geotropism. In the shoots, the auxin causes more growth on the underside, which does not receive any light, so the shoot bends upwards towards the light. This is called positive phototropism.

Pages 78–79
QUICK TEST
1. To make enzymes used in respiration and photosynthesis; **2.** Poor root growth, purple younger leaves; **3.** To carry water and dissolved minerals from the roots to the shoots and leaves.
EXAM PRACTICE
1. Any two from: Elongation/enlargement in plants is the main way plants gain height; Animals stop growing, plants grow continually; Cell division in plants only takes place at the root and shoot tips; Plant cells retain the ability to differentiate throughout life; Animal cells lose this ability. **(2)**